Housing and social theory

T0251095

Housing and social theory

Jim Kemeny

Routledge
Taylor & Francis Group

LONDON AND NEW YORK

First published 1992
by Routledge
2 Park Square, Milton Park, Abingdon, Oxfordshire OX14 4RN

Simultaneously published in the USA and Canada
by Routledge
711 Third Avenue, New York, NY 10017

Reprinted 1999, 2001

Transferred to Digital Printing 2007

Routledge is an imprint of the Taylor and Francis Group, an informa business

First issued in paperback 2015

© 1992 Jim Kemeny

Typeset in Times by LaserScript Limited, Mitcham, Surrey

All rights reserved. No part of this book may be reprinted or
reproduced or utilized in any form or by any electronic,
mechanical, or other means, now known or hereafter invented,
including photocopying and recording, or in any information
storage or retrieval system, without permission in writing from
the publishers.

British Library Cataloguing in Publication Data
Kemeny, Jim
   Housing and social theory / Jim Kemeny
      p.   cm.
   Includes bibliographical references and index
   1. Housing–Sociological aspects. 2. Social structure.
I. Title
HD7287.5.K46   1991
363.5–dc20   91–11123
                                                                  CIP

Library of Congress Cataloging-in-Publication Data
   Has been applied for

ISBN 978-0-415-06273-2 (hbk)
ISBN 978-1-138-97201-8 (pbk)

Publisher's Note
The publisher has gone to great lengths to ensure the quality
of this reprint but points out that some imperfections in
the original may be apparent

For Martin, my son and friend

Do not allow public issues as they are officially formulated, or troubles as they are privately felt, to determine the problems that you take up for study. Above all, do not give up your moral and political autonomy by accepting in somebody else's terms the illiberal practicality of the bureaucratic ethos or the liberal practicality of the moral scatter. Know that many personal troubles cannot be solved merely as troubles, but must be understood in terms of public issues – and in terms of the problems of history-making. Know that the human meaning of public issues must be revealed by relating them to personal troubles – and to the problems of the individual life. Know that the problems of social science, when adequately formulated, must include both troubles and issues, both biography and history, and the range of their intricate relations. Within that range the life of the individual and the making of societies occur; and within that range the sociological imagination has its chance to make a difference in the quality of human life in our time.

C. Wright Mills, 'On intellectual craftsmanship' in Mills (1959: 248)

Contents

Figures

Preface and acknowledgements

The origins of this book lie in the mounting frustration I felt with the way in which an earlier book, *The Myth of Home Ownership* (Kemeny, 1981), was being received. All of the attention was focused on the tenure issue, and the underlying divergence thesis which tenure was meant both to illustrate and to help to explain was ignored. It was the growing interest in theory among housing researchers during the 1980s that ultimately led to my realisation of what was wrong with that book. Catalytic in this regard was the work of my colleague, Stephan Schmidt, who soon after moving to the National Swedish Building Research Institute read my book and in a spirit of healthy scepticism determined to test the relationship between tenure and the welfare state using variable analysis, the results of which (Schmidt, 1989a), somewhat to his surprise, confirmed my findings.

This encouraged me to re-address the issue by positing an explicit divergence thesis in a work that was intended to be more theoretically developed. The immediate result of this was a long paper presented at seminars, first at the Institute and then at the School for Advanced Urban Studies in Bristol. That paper subsequently became three papers, which were later drastically reworked to become chapters 1, 4, and 7 of this book. I then realised that I needed to do more work on the theoretical and epistemological issues underlying the thesis of the new book, and that as a result the question of the divergence thesis would take up only a modest portion of the total study. In short, I realised that the book was becoming unbalanced, and that a prior book was in the process of emerging, addressing the more fundamental issue of the theoretical status of housing studies and using the divergence thesis as just an illustration of this. It therefore turned out very differently from originally envisaged and I have ended up writing a prior piece of work, leaving the original project still to do.

The book was largely written in Britain between June 1989 and May 1990. However, I was funded throughout by the Institute, and it would not have been possible to write it without the Institute's support. For the first part of the period I was based at the Department of Sociology at the University of Aberdeen. In the autumn of 1989 I moved to the School for Advanced Urban Studies in Bristol and in Spring 1990 to the Centre for Urban and Regional Research at the University of Sussex. I am grateful to all three for the provision of facilities that enabled this book to be written.

I should also like to thank colleagues at these places for commenting on drafts of chapters and their interest in my work, particularly Mark Shucksmith, Norman Stockman and Nick Williams at Aberdeen; Craig Gurney, Ray Forrest and Alan Murie at Bristol; James Barlow, Peter Dickens, Simon Duncan and Peter Saunders at Sussex. Others who have kindly commented on draft chapters have been Mike Harloe at Essex University, Walter Matznetter at Vienna University, Steve Merrett at the Bartlett School of Architecture in London, Hannu Ruonavaara at Turku University in Finland, and Bo Bengtsson, Lars Nord and Eva Sandstedt at the National Swedish Institute for Building Research. Stephan Schmidt and Ray Forrest both read and commented on the penultimate draft in its entirety, and their comments led to extensive redrafting of parts of it.

At the Institute I benefited greatly from discussions with colleagues, too many to name individually. Most of all I am indebted to Stephan Schmidt, with whom I have had numerous and sometimes extended discussions which have been invaluable in helping me to explicate my ideas. As a result I sometimes found it difficult to separate his ideas from mine and I crave indulgence for any unintended plagiarism that may have resulted. In addition, participants at the Sociology Seminar held at the Institute in October 1990 on the penultimate draft of Chapter 8 provided many constructive and helpful comments.

Last but not least I should like to acknowledge the superb technical and material support of the Institute's research facilities and, even more importantly, I should like to thank the Institute's administrative staff for invaluable support. In particular, Hasse Johansson's understanding and patience made possible my extended stay in Britain, the library staff provided excellent inter-library loan facilities, and secretaries Birgitta Bäck and Gunilla Bloom did much prompt and efficient computer and formatting work on the text and figures.

Introduction

Housing research has traditionally been concerned with measuring the extent of housing shortages and specifying its dimensions, for example in terms of overcrowding, poor sanitation, and a shortfall of dwellings in relation to households. The latter, in particular, has eclipsed issues of overcrowding and sanitation in the post-war period. Estimates of the numbers of households and the numbers of dwellings, as well as the number of 'hidden' households resulting from delayed household fission in response to housing shortages, comprise a key measure of total dwelling shortfall and the amount of new construction that is needed to overcome it, and have traditionally constituted the starting point for housing research (Cullingworth, 1960: ch. 1; Donnison, 1967: 33–5). This practical and policy-oriented approach to housing effectively defines housing in terms of dwelling units, or what is sometimes termed 'shelter' (Abrams, 1964).

However, since the publication of Rex and Moore (1967) in which the first genuine housing concept – that of 'housing classes' – was introduced, housing studies has been undergoing a slow but certain metamorphosis as growing conceptual and theoretical awareness has gradually begun to permeate research. The process was given considerable stimulus by the growth of Marxist theory, and particularly Althusserian structuralism, during the 1970s, followed by a Weberian reaction during the 1980s. As the 1990s begin, after nearly a quarter of a century of slow but sure development, the pace of change promises to quicken dramatically.

In spite of this, the theoretical development of housing research remains rudimentary and leaves much to be desired. A central problem of much of housing studies is that it retains a myopic and narrow focus on housing policy and housing markets, and neglects broader issues. Housing studies is still far too isolated from debates and theories in the other social sciences and what is needed now is further integration into these.

But, as I argue in another context later on, changes in ideas cannot be understood outside of changes in social structure. Since the mid-1980s an institutional infrastructure for teaching and research in housing has been developing in many European countries with almost breath-taking speed. Research and teaching centres are being established, chairs in housing are being founded, international academic journals in housing, using the peer review system, have been started, regular international housing conferences have been instituted, and most recently the European Network for Housing Research has been established. In a remarkably short time, we are witnessing the establishment of housing studies as a new academic field with its own power structures of agenda-setting and gatekeeping: over making appointments to teaching and research posts, defining the content of courses, deciding which research proposals are and are not accepted for funding, deciding which papers become journal articles and which books get to be reviewed and by whom, and so on.

This development has its dangers. Institutions can have a sclerotic effect on research, hardening perspectives into 'schools' and handing down traditions which are not easy to challenge, let alone change or bypass. Even more worrying is the very real possibility that institutionalisation will accentuate a still chronic degree of isolation of housing from broader social concerns. The institutionalisation of housing studies could easily legitimise the field as a narrow specialism that can be treated out of the context of society as a whole and in abstraction from broader social science debates and controversies. My concern is that the institutionalisation of power structures in housing studies is taking place too soon, and that there is a real danger that abstracted empiricism and policy-determined research will become entrenched, thus stifling, or at least retarding, the emergence of critical and reflexive housing research.

The urgency and importance of this task cannot be overestimated. Housing is one of those subject-determined fields that all too easily becomes an unreflexive empirical study in abstraction from society as a whole. It shares this characteristic with many other 'fields' such as education, transport, and health. Some of these fields, and most notably that of education, have become institutionalised and ossified into an underconceptualised backwater of the social sciences. Others, such as health, are now beginning to free themselves of traditional limitations, and to develop a much broader and theorised understanding of their focus and place in the social sciences. The fate of housing studies could be dire if it moves into its phase of intense institutionalisation without a deep and thorough-going evaluation of the grounds of the subject.

This book is therefore very much a product of its time. Its purpose is to further the theoretical development of housing research. The most urgent task in this respect is the need to clarify the status of housing studies as a field in relation to other social sciences.

Thus if there is one dominant theme in this book, it is the need consciously to re-integrate housing into broader issues of social structure and to do so by explicitly relating housing to debates in the social sciences. Such a task has a number of components. The first requirement is that we begin to discuss the relationship of the field to other fields and disciplines: to begin to define the field. So far little or no interest has been shown in the question of just what it is we are studying, and why. This question is addressed in the first chapter.

The second requirement is that we begin to develop a greater sense of reflexivity, by examining more closely the epistemological grounds of what we are doing: the substantive focus of housing research, the methods we use, the data sources we tend to take for granted, the questions we pose, and the received wisdoms, often from policy-makers, about what does and does not constitute 'a housing problem'. These make up the subject of Chapter 2.

The first two chapters of the book therefore comprise a wide-ranging discussion of broad issues of the nature of housing studies. The second part of the book takes up the theme concerning the need to integrate housing into other disciplines by examining the ways in which debates in the disciplines can contribute to a richer and deeper understanding of housing issues. I am concerned to show how, by neglecting debates in other social sciences, housing studies is thereby unnecessarily theoretically impoverished. I choose three issues which relate closely to housing research: theorising the state in political science (Chapter 3), theories of social change (Chapter 4), and theories of welfare (Chapter 5). In the first two of these, my approach has been to present key debates and then to critique the housing literature to show how its neglect of them has impoverished it. The third issue is treated somewhat differently, and rather than repeat the critique exercise yet again I try to relate social administration to the concept of 'residence'.

The third part of the book must be seen as an application of the ideas developed earlier, which takes the form of a case study in theorising housing. I attempt to develop a divergence thesis, building upon my earlier comparative research. In doing so I draw on a range of theories. My aim here is twofold. It is first intended as a development of my earlier undertheorised comparative research, though I make no claim to present a fully developed divergence thesis.

A second – and in one sense more important – aim is to illustrate how I envisage theorised housing research to be conducted. I do not claim that either the subject matter presented in Chapter 8 or the particular theories I choose to apply to it are in some absolute sense 'correct' or superior to others. I could equally well have chosen to illustrate my views of a theorised housing studies by positing a diametrically opposite thesis, such as convergence, and drawn upon the theories of Rostow, Kerr, Bell, or any other major social scientist in order to understand and explain convergence.

However, I do claim that certain important principles concerning the relationship between housing and other dimensions of social structure are enduring and comprise a fundamental tension that all housing researchers must address. These concern the epistemological issues raised in Chapter 2. The last chapter of the book therefore returns to these questions. The task I set myself is to sketch the outlines of a conceptual approach to housing within the sociological tradition. In particular, this concerns the embeddedness of housing in social structure and its far reaching impact upon it. I argue that housing comprises such a fundamental and major dimension of social structure that the study of residence can and perhaps should be the central contribution of housing studies to our understanding of society. My task in this last chapter is therefore to assemble some of the key elements of a sociology of residence which I would see as the heir to the moribund sociology of housing. For this reason, I would subtitle the book 'a prolegomenon to a sociology of residence'.

Over and above this major task, an important principle that permeates the book is that of reflexivity, and the need for a *critical* approach to housing studies. This is such an important precondition for the development of a sociology of residence that it bears repeating, by way of conclusion. Theorising is not merely the mechanical application of ideas from one field to another. It is essentially the use of what C. Wright Mills called 'the sociological imagination': calling into question aspects of social structure that tend to be taken for granted and using fantasy and lateral thinking in the craft of sociology to solve problems, often reformulated in novel ways. It also involves the vital ingredient of emotional involvement. Science is not a neutral and purely cerebral exercise. It must rather be understood as a socially embedded act in which involvement and detachment interfold in complex ways.

Housing studies has never developed true reflexivity and autocritique, precisely because it has remained largely outside the major intellectual developments in the social sciences during the postwar period. It was never influenced by structural functionalism which dominated the social sciences during the 1940s and 1950s. By the same token, it was never touched by the reactions to structural functionalism that took place during the intellectual upheavals of the 1960s. These were manifested in the emergence of critical social science in the tradition of Wright Mills and in the development of symbolic interactionism and ethnomethodology. These were fundamental non-Marxist critiques of overdeterminism and moral conservatism that never found an echo in housing studies. The field therefore never had the opportunity to engage in the intense self-critique and questioning of fundamental directions that took place in sociology and other social sciences. Such a phase in development is an essential prerequisite, I would argue, to developing theory in housing research. And since it never happened – though there are signs that this is beginning to change – it may be necessary for housing researchers to retrace their steps and begin to engage in some of these issues that are old hat in most of the social sciences.

The tasks I have set myself in this book are too ambitious to do more than provide initial thoughts on a theorised housing studies. Others will have to build further, developing fruitful lines and abandoning others. But if the book succeeds in stimulating debate over central issues of housing studies it will have achieved its purpose.

Part I
Housing and metatheory

1 The disciplinary basis of housing studies

INTRODUCTION

During the early postwar period, housing research was carried out largely within established disciplines such as economics or sociology, or in social administration or social work departments, and there were no academic journals devoted to the field. Today housing is emerging as a specialist field and the rapid institutionalisation of housing research provides the context for an evaluation of the relationship of housing studies to older, more established cognate fields such as urban studies and to the social science disciplines from which housing researchers often draw inspiration and concepts. It prompts the question of what exactly is this field of housing studies? And how can we begin to understand the place of housing research in the context of wider social questions, and in relation to other areas of the social sciences?

In this chapter I propose to conduct a general and fairly wide-ranging examination of the place of housing studies in the social sciences. My concern is both to maximise the benefit of advances in other areas of social science and to contribute to wider debates outside of housing issues narrowly defined. However, before we can consider the place of housing studies in the social sciences it is necessary to clarify what is meant by 'disciplines' and the sense in which the term is used in this discussion. I largely limit my observations to housing research in English-speaking countries, though my direct knowledge of housing research in Scandinavian countries and what I know of housing research from translated work in other countries suggest that similar principles are likely to apply everywhere.

ON THE NATURE OF DISCIPLINES

In general, the social science disciplines can be seen to be based on dividing the social world into a number of dimensions. Sociology, for example, 'dimensions out' social relationships which are often conceptualised in terms of the abstraction known as social structure. Economics does the same for the market. Psychology dimensions out individual mental processes. Political science dimensions out power and political institutions. Geography dimensions out space; history dimensions out time, and so on.

Each discipline develops its own sets of conceptual tools for the analysis of its particular dimension. Theories are explicated and tested, and a characteristic mode of discourse is evolved through the generations, with its own major debates and controversies. The point about this is that each discipline is based on researchers being 'disciplined' into thinking in certain ways and in critically evaluating existing theories and concepts developed by others within that mode of discourse.

Disciplines are based on a process of conceptual abstraction. That abstraction provides the epistemological basis for the discipline and provides it with a selective frame of analysis. Disciplines are not normally defined in terms of a concrete field or subject of analysis (though, as we shall see, they may be). They are more usually defined by a frame of reference, even if some frames of reference prove in practice to be more amenable to theorising than others.

A good example is the sociological frame of reference, which, by abstracting out such a general dimension as social structure, provides wide scope for theorising. Geography, by contrast, appears to enjoy less scope for theorising since the focus upon the concept of spatial relationships is narrower and more restricted. The same is true for history in relation to temporal factors. The disciplines based on time and space within the social sciences are therefore much less theoretical and more empirical in nature, largely because the scope for theorising time and space as dimensions is limited. So although historians and geographers have attempted to theorise temporal and spatial dimensions respectively of social phenomena, the vast bulk of work in these disciplines has been devoted to describing social phenomena in terms of temporal drift and spatial configuration. In sociology, by contrast, much more effort has gone into developing the theory of social structures.

Having said this, however, it is important not to exaggerate the extent to which disciplines are logical and rational structures of thought. Not all disciplines are dimension-based, and some subject areas have succeeded in becoming established as disciplines: social work and social administration are examples of this. Disciplines are to a large extent the product of power struggles taking place both within universities and between the research world and funders: including, crucially, the state. Care must therefore be taken not to reify the concept of a discipline into a theoretically pure phenomenon. It so happens that the discipline I am most concerned with in the context of comparative housing studies and the subject of theories of social change – sociology – has a long tradition of theoretical work and a wealth of concepts to draw upon.

What, then, is the basis of housing studies? It is neither a discipline in the sense that it abstracts out a dimension of society, nor is it an established 'subject-based discipline' in university power structures, even if it is rapidly becoming so. Before addressing this question it might prove instructive to consider briefly two closely related fields: urban studies and social administration.

THE CASE OF URBAN STUDIES

The growing interest in theory within urban studies has led to considerable effort being made to identify the epistemological grounds of the field. The question of whether it is possible to theorise the urban has been considered in some detail by Saunders (1986). From a wide-ranging overview of the attempt to theorise the urban by major social theorists since Marx, Saunders concludes that the urban does not provide the basis for special theoretical focus, and that all previous attempts to find one have failed.

Urban studies is a problematic field because it is based on selecting out a dichotomous element in social structure, namely the 'urban' *contra* the 'rural', a fact recognised by Frankenberg (1966) in his concept of the 'urban–rural continuum'. At the same time, urban areas are an appealing focus for research because towns and cities in industrial societies appear to have concentrated in them many of the major social problems of modern society. More important, perhaps, is the underlying belief that the urban constitutes the essence of 'modernity' and that it is in the urban that the basic dynamic of social change can perhaps be found. There is therefore considerable interest in developing epistemological grounds for theorising the urban.

The search for an alternative way of conceptualising problems involving an urban dimension has continued. Perhaps the most convincing recent attempt has been that of Gregory and Urry (1985) in their reconceptualisation of regional issues by identifying the interface between social structures and spatial factors as constituting a theoretical focus. Such a redefinition would abolish the urban as a focus for theorising, but would integrate urban and rural into a regional studies based on the interaction between spatial and social dimensions (geography and sociology).

Gregory and Urry bring together a number of papers which, when taken together, represent different ways of integrating spatial factors into social theory. They are critical of the aspatial nature of social theory, and argue that an integration of human geography and social theory provides the basis for a new and more comprehensive approach to explanation in the social sciences. As they put it:

> The aim of this book is to minimise some of the academic space between human geography and social theory in order to establish a new agenda for theoretical and empirical work and so explore new and challenging 'common ground'.
>
> (Gregory and Urry, 1985: 8)

However, the book is a highly programmatic statement. It provides no real basis for a new perspective, but rather signals its desirability. It is too early to say whether approaches of this nature are likely to result in the emergence of a new socio-spatial perspective in which urban and rural issues share the same epistemology. There is certainly considerable work now being done on issues of 'locality' (see Duncan and Goodwin, 1988) which highlights the influence of spatial distribution on social structure and which could ultimately bridge the gap between urban and rural studies. Even more promising is a recent attempt to develop an integrated socio-spatial approach to restructuring in locality studies that explains spatial change in terms of social, cultural, and political processes (Baggueley *et al.*, 1990). But all this is very tentative and it will probably be some time before it is clear whether or not a new perspective is emerging.

THE CASE OF SOCIAL ADMINISTRATION

Housing studies has a close – even intimate – relationship to social administration. This is because housing is itself an important area of concern to social administration and because much housing research has its origins in the research tradition of social administration. However, social administration has undergone something of a transformation since the early 1970s, with a rapidly growing theoretical awareness greatly exceeding that within housing studies (Forder *et al.*, 1984; George and Wilding, 1976; Gough, 1979; Mishra, 1981; Offe, 1984; Pinker, 1971; Taylor-Gooby and Dale, 1981). The kinds of issues that are grappled with in this book in terms of housing have been the subject of debate in terms of welfare within social administration since at least the mid-1970s. Part of that debate concerns the relationship of social administration to other subjects, to theory, and to the social science disciplines.

Mishra (1981: ch. 1) has delineated the principal characteristics of social administration: its national British focus, values of interventionism and piecemeal reform, supra-disciplinary (or field) orientation, and empiricism. This list will have a familiar ring to it for housing researchers. With the partial exception of the ethnocentric focus on the British welfare state – which has anyway changed during the 1980s towards a much more international comparative focus in both housing and social administration – Mishra's depiction of social administration also applies to much of housing studies.

Mishra summarises succinctly the principal dissatisfactions that have emerged over recent years with the nature of social administration as a research area. He also points out that there is no consensus over the relationship between social administration and the social science disciplines: for example, Donnison defining it as simply a 'field' drawing on various disciplines while Titmuss saw social administration as an emerging synthetic discipline in its own right (Mishra, 1981: 20).

This latter view is endorsed by Carrier and Kendall (1977). They attempt to define the subject matter of social administration in terms of 'welfare activities whose manifest purpose is to influence differential "command over resources" according to some criteria of need' (Carrier and Kendall, 1977: 27). This definition is broader and more general than traditional definitions in terms of statutory welfare, such as put forward by Titmuss, for example, in that it focuses on the distinction between private market and welfare provision in its widest sense.

But even this distinction does not provide the basis for a new synthetic discipline of social administration, and an even more general and wide-ranging approach is in the process of emerging: one in which welfare is not seen as limited to a public–private distinction but is defined as a basic characteristic which can take many forms and which constitutes the subject focus of social administration. This approach is emerging out of the debate over the supposed monopoly role of the state in welfare provision: a position that has come under increasing attack in recent years. The growing importance of other forms of provision – by the market, employers, voluntary agencies, and informal networks, sometimes known as 'welfare pluralism' (Johnson, 1987) – has if nothing else drawn attention to the need for a broader definition of welfare. Rose (1986a, 1986b) argues that welfare is provided in many forms but principally by households informally, by the state, and by the market, and that the 'welfare mix' of these three types of welfare varies both over time and between countries.

This point will be returned to in Chapter 5. For now we need only note that social administration is gradually evolving a conceptual approach to the subject of welfare. The process is still in its early stages but it is clear that, by moving away from a narrow statutory definition of welfare in which existing practices and laws define and delimit what is and is not a legitimate subject for social administration, the way is opened up to develop a definition of welfare in terms of a more general social dimension.

What form this will take remains as yet unclear. But one possibility could be to base it on the Marxist concept of the reproduction of labour, involving some concept of 'mutual aid', irrespective of the providing agency, whether it be state, employer, voluntary, private, neighbourhood, or family.

SUBSTANTIVE FOCUSES: HOUSING, HOME, AND RESIDENCE

The cases of urban studies and social administration suggest that both in different ways have the potential to develop conceptualised dimension-based approaches to the respective subjects. Of the two, social administration is the most promising and is in some ways more relevant as a model for the conceptual development of housing studies. The solution of attempting to develop a disciplinary focus in terms of the interface between space and society, which may well work for regional studies, is clearly inappropriate in the case of housing studies. 'Housing' is a substantive focus – a focus upon dwellings; it is not one pole of a dichotomous concept, as is an urban focus, and so cannot be integrated with a polar opposite in the way that urban and rural dimensions can be combined to create a socio-spatial regional dimension. It would seem more appropriate to develop a conceptual basis for housing by refining the concept of housing in a parallel manner to the refinement that has been taking place of the concept of welfare in social administration. How might this be done?

Housing studies is clearly about housing. But this tells us little. Housing, after all, in its simplest and crudest sense, is the bricks and mortar or other building materials that comprise the constructions within which people live. But as a field within the social sciences, housing research equally clearly involves the examination of the social, economical, political and other relationships that centre on housing. We might, therefore, by way of providing a starting point, provisionally define housing studies as the study of the social, political, economic, cultural and other institutions and relationships that constitute the provision and utilisation of dwellings.

This amounts to conceptualising housing in terms of that dimension of society concerned with 'shelter', as is sometimes done (Abrams, 1964). Such an approach would define housing studies as the focus for the social relations that directly and indirectly involve the activities of planning, constructing, managing, and use of shelter. Yet this is hardly satisfactory since housing is so much more than shelter. Indeed, it is the very narrowness of a 'bricks and mortar' approach to housing that needs to be avoided. It is no coincidence that housing studies is so called. It reflects the planning and social administrative origins of the field, and it is precisely this focus that needs to be modified, or at least broadened.

There is clearly dissatisfaction with the focus of housing studies. One attempt to broaden the scope of housing studies has been that of Saunders and Williams (1988) who want to redefine housing issues in terms of 'the home', thereby focusing upon the household – rather than the dwelling – and the social processes that are associated with it. However, Saunders and Williams' argument explicitly comprises part of a wider concern to focus upon consumption rather than provision issues. This is a complicating factor in their concern to put the home at the top of a research agenda which has more to do with substantive issues than epistemological ones.

Moreover, the shift in focus away from dwellings and towards the households that inhabit them, which is a consequence of a focus upon the home, avoids rather than resolves the conceptual ambiguities surrounding the relationship between dwelling and household. These ambiguities are reflected in the way in which the words 'house' and 'home' are often used interchangeably or are closely coupled as in 'house-and-home'. I have argued elsewhere that the concepts of 'household' and 'dwelling', basic as they are to housing researchers, are confusing and unclear because they are each defined in part in relation to the other (Kemeny, 1984). This issue lies at the heart of the problem of what constitutes the substantive focus of housing as a research field and is a recurring theme throughout the book.

Focusing upon the home therefore unnecessarily limits the scope of housing research. A broader concern is desirable; one which embraces locational factors and ties housing studies into macro issues of the nature of social structure. If there is any one dimension of social structure that is central to the way in which it is organised, it is housing. Housing comprises such a major aspect of the organisation of daily existence that it very naturally acts as a focus for the study of a large number of social issues, and particularly those relating to comparative social structures. This can be simply illustrated by taking two major dimensions of housing and showing how they affect social structure: the spatial organisation of housing, and the way in which households pay for it.

The spatial impact of housing is most clearly demonstrated in terms of the impact of dwelling-type on urban form. It makes an enormous difference, far beyond the narrow issue of shelter, whether urban areas are predominantly made up of detached houses or high-rise flats. The knock-on effects will be major, if not determinant, on, for example, the organisation of urban transport which in turn will affect profoundly patterns of sociability, uses of public and private space, and differential accessibility by such dimensions as age, gender, and class. Other substantive bricks-and-mortar focuses – for example, the way in which school buildings are organised, or the spatial organisation of hospitals and other forms of medical care – also have some impact on social structure but they are much less profound and far reaching than different ways of spatially organising housing.

The other dimension – the economic organisation of housing – is equally far reaching in its consequences. The main way in which this is manifested is in terms of different forms of paying for housing by households. The most obvious and dramatic difference in this respect is that between owner occupation and renting, which, under normal financing arrangements, dramatically affects the manner in which housing costs are paid over the household life-cycle. It is clearly of primary importance to patterns of consumption if housing costs are concentrated at the beginning of the family cycle, as they are in owner occupation, or spread out over the whole cycle fairly evenly, as they are in renting. The organisation of housing finance and the extent of owner occupation in different social groups is therefore of major importance to spending patterns at different ages and among different social groups, as well as having considerable impact on the extent of resistance to taxation.

Both of these dimensions of housing – its major spatial effects on the social organisation of urban areas, and its high cost as a percentage of total household expenditure – combine to give housing a uniquely important place in the analysis of social structure. This is a fact of which housing researchers are clearly aware. However, awareness is one thing; understanding the theoretical – as against the practical – implications of this and organising research in those terms is another.

It is therefore particularly handicapping that housing remains so under-conceptualised and that we have barely begun to think about a definition of housing that relates it to social structure. This is why some concept of 'residence' would be useful as a starting point. It would begin to focus attention on the wider social structural relationships within which narrow issues of shelter are embedded, but which currently tend to get ignored or lost. Housing is not just about dwellings, nor even just about the households that live in them. Nor is it limited to the interaction between households and dwellings in, for example, the home. It also includes the wider social implications of housing. A case could therefore be made out for reconceptualising housing in terms of 'residence' to capture its broader social issues.

'Residence' as a concept directs attention to the dwelling as home, *but within its locational context*. A peculiarity of the English language is that the word 'residence' has become something of an anachronism in everyday usage. The word is still used in some research contexts (as in 'residential mobility'), but in everyday terms the verb 'to live' has come to add to its meaning of 'to be alive' that of 'to reside'. Many languages distinguish in everyday talk between 'residing' and 'living'. It is not possible, as it is in English, to say that one 'lives' in a town when one really means 'resides' in it. This lapsed use of the word 'residence' in English is unfortunate, because 'residence' carries many social connotations, particularly concerning the ways in which individuals or households tie into wider circles of locality, and beyond.

The concept of residence could therefore usefully form the substantive focus of housing studies in place of either 'dwelling' or 'home', to provide a broader contextual perspective. The central feature of a focus upon residence as the basis for housing studies must be its clear social structural orientation. It embeds housing issues in their locational and wider social contexts, thereby shifting the emphasis away from the bricks and mortar implications of 'housing' towards social dimensions. It highlights the social organisation of housing, both in terms of its location and its form, and the interplay between these. It includes issues of urban form, limitations which it places on restructuring, the relationship between housing and welfare, and the entire gamut of institutional and organisational arrangements that impinge directly and indirectly on issues of residence. Residence as a dimension of social structure can thereby be seen as a key dimension of the social organisation of modern society. It also consitutes an intermediate, meso, level between large-scale macro societal processes and individual level micro interaction.

INTERDISCIPLINISM, HOUSING AND SOCIAL THEORY

In a review of the disciplinary basis of social medicine, Turner (1990) divides academic specialisms into a hierarchy of increasing complexity, with disciplines combining at a simple level into *multi*-disciplinary studies in which different disciplines are drawn on in an *ad hoc* manner and then, more interactively, into *inter*-disciplinary studies in which theories and approaches from different disciplines are integrated with one another in innovative ways.

The atheoreticity of housing studies suggests that it is characterised by the dominance of a multidisciplinary approach, drawing on several disciplines but not integrating them with one another in new ways. This is a useful starting point but an oversimplification. I want to argue here that in practice, although deriving from multidisciplinary interests, housing studies has created its own adisciplinary or nondisciplinary mode of discourse.

A genuine interdisciplinism requires a prior firm anchoring in the individual disciplines before any adequate interactive analysis across disciplines can be developed. This suggests an initial widespread splintering of the field into a number of purely discipline-based housing studies: a housing sociology, a housing economics, a housing geography, a housing ethnography, and so on, each with its intellectual basis within its own parent discipline and oriented to the problems, theories, and debates within the discipline. Only then would it be possible to develop a genuine interdisciplinary housing studies. There are, then, two contrasting moments: a return to individual disciplines, followed after a period of development of disciplinary perspectives, by a process of synthesis in which new integrative approaches and theories are developed. Each is an essential component in the epistemological construction of housing studies as an integrated field. The end result should then provide a clearly interdisciplinary basis but – and this is crucial – one that is derived from explicit disciplinary concerns.

The extent to which housing research from within different disciplines is carried out with reference to internal disciplinary issues rather than with reference to researchers from other disciplines varies between disciplines within the field of housing studies. But most housing research is not discipline-oriented. There is a remarkable lack of grounding of housing research in the social sciences.

The isolation of housing studies from the major debates in the social sciences is well illustrated by the case of housing and sociology, but could equally well be illustrated from other disciplines, notably political science or anthropology. Sociology underwent an explosive growth during the 1960s and early 1970s which contributed to the flowering of theoretical and conceptual work in the discipline. Very few of these debates are reflected in housing studies. The widespread influence of Parsonian structural functionalism during the 1950s was not reflected in housing research. Nor was the reaction to this that took place in the 1960s, with the emergence of symbolic interactionism and the re-emergence of Marxism. Nor were the methodological debates of these years, notably over qualitative versus quantitative research, which was rife in sociology, at all evident in housing studies. In particular, the extensive critical literature on the use of official statistics in sociology would have been directly relevant to housing researchers.

But what about the major debates in the social sciences that have a direct bearing on housing issues as defined by housing researchers themselves? Astonishingly enough, the same is true of these. Two of the most glaring examples are probably the neglect by housing researchers of theories of state power and theories of social change. Both of these will be considered in more detail in Chapters 3 and 4 respectively. But most housing researchers have neither been interested in drawing upon the ideas of sociologists and political scientists in these terms, nor in critically testing them in their own field studies. Despite growing interest in theoretical issues there remains a strong tendency for housing researchers to bury themselves in their own empirical and policy issues with almost complete disinterest in such 'abstract' questions.

The extent to which housing studies has ignored the larger debates within sociology and political science is on reflection quite amazing. How is it possible that the enormous and stimulating outpouring of early postwar social science research made little or no impression on housing researchers, and why, in the face of such a flowering of the social sciences, did housing studies remain an intellectual backwater?

The reason would seem to be that debates in housing studies have been nested within issues deriving from the field itself and oriented towards internal debates. This is precisely the danger with multidisciplinism. It too easily becomes nondisciplinary, or adisciplinary, resulting literally in a lack of discipline in the organisation of research. Instead of handling housing issues within established disciplinary modes of discourse, the temptation is to pick a little here and a little there from different disciplines and to avoid having to draw on or even acknowledge established debates, concepts and theories in any discipline. Instead, the social construction of housing studies takes place around empirical and policy issues in abstraction from theories of any kind. The analysis thereby gravitates to a mode of discourse that represents the lowest common denominator to the social sciences with a resultant focus upon policy and empirical work and a neglect of theory (Kemeny, 1988). The end result of this subject-fixated approach is that abstracting 'housing' out of social structure and focusing upon it leads to a failure to integrate it into the wider social processes of which it is a part. Housing studies therefore tends to become a specialism, divorced from wider social issues. It becomes a sterile and limited empirical focus, concentrating on analysing the housing market and housing policy.

My own experience as a housing researcher reflects the way in which this takes place at the individual level. I came to housing studies after many years as a general sociologist with a long-standing interest in questions of social power and the debates and issues surrounding pluralism, élitism, ruling class theory, and corporatism. Yet I made no attempt to use this knowledge in my housing research. Instead, I naturally slipped into what might be termed 'the dominant mode of discourse' within housing studies, and subconsciously pigeon-holed my sociological experiences as not relevant. This, I suspect, is the case with many housing researchers who have discipline-based social science backgrounds. It is all too easy to recognise that housing researchers with different disciplinary backgrounds to oneself are unlikely to be aware of the debates that have taken place within one's own discipline. Thus, for example, a political scientist, rather than going into the intricacies of the Miliband–Poulantsas debates on political power and the state for the benefit of a housing researcher with, say, a geography discipline background, finds that it is simpler to reduce to the lowest common denominator and leave aside specialist disciplinary issues. This is just another way of saying that housing studies has developed its own discipline-neutral mode of discourse which dominates the literature, and into which researchers from different disciplines have a tendency to slip almost unawares.

One of the consequences of this 'debate-blindness' and lack of awareness of existing literature is that apparently new issues are sometimes developed into major debates even though they have in fact taken place outside of housing studies, sometimes long ago. Some of the eureka-like 'discoveries' that have taken place in housing and urban studies therefore have a *déjà vu* quality about them to those familiar with such issues from a discipline in another context.

Two recent examples of this in urban and housing research have been the interest in 'realist' methodology, associated with the work of the philosopher Bhaskar (1975), and the rediscovery of the individual in urban studies, as traced by Thrift (1983). These appear to be akin to the reinvention of the wheel, since both have extensive literatures which could have been drawn on. In the former, the vast qualitative methodology literature deals with the insights of realism, going back at least to Cicourel (1967), and Glaser and Strauss (1967). The latter, in particular, and the concept of 'grounded theory' appears to be very closely related to the concept of 'realism'.

In the second example, that of the rediscovery of the individual in urban studies, there are many microsociological approaches dating back to the early 1960s if not before that could be drawn upon, such as exchange theory, reference group theory, and symbolic interactionism, not to mention the extensive micro–macro literature that has emerged since the mid-1970s (Collins, 1981; Knorr-Cetina and Cicourel, 1981). The reason why none of these appears to have gained ground in urban studies seems to be that one of the leading theorists in this field – Giddens (1979) – happened to be drawn into issues of space as a result of contact with geographers of the Swedish school of time-geography (Carlstein, 1981; Hägerstrand, 1970, 1974; Lenntorp, 1976). As a result of this, his work was brought to the attention of urban geographers. It so happens that Giddens is very dismissive of microsociological perspectives – and in particular of symbolic interactionism – in developing his own theory of agent–structure dynamics (Giddens, 1979). But the result of this particular citation network has been the great prominence given to Giddens' highly programmatic theory of structuration and the little or no awareness of the large existing literature on the agent-structure problematic in sociology, particularly the work of Randall Collins and of symbolic interactionists (for a discussion of this see Kemeny, 1987b).

No doubt other examples could be found in reverse, where urban and housing sociologists reinvent wheels long existing in other disciplines, or struggle to develop concepts in apparent unawareness of the prior existence of a substantial body of literature which could usefully be taken as a starting point, however inadequate and underdeveloped it might be.

All this is not to dismiss multidisciplinary work as inherently inferior. Rather, it is to point to the dangers of multidisciplinism slipping into nondisciplinism and abstracted housing empiricism, and losing touch with the mainstream disciplinary concerns within the social sciences. Multidisciplinism within housing studies has become 'nondisciplinism' because of the prevalent tendency to fail to work out of clearly discipline-based concerns by taking the issues and theories of one's own discipline and linking them into housing issues. Multidisciplinary housing studies must therefore be very firmly grounded in debates from the disciplines in order to provide the theoretical foundation which enables concepts and theories from different disciplines to be fruitfully combined.

This is a fate that befalls many specialisms with a subject-based focus. Turner (1987), for example, argues against the way in which medical sociology has developed in precisely these terms. Medical sociology deals with the interweaving of issues of health and sickness with other aspects of social structure. Turner argues that the field has become too narrowly focused upon the study of medical institutions and practices: the aetiology of disease, doctor–patient relationships, hospital organisation, health service provision, *etc.* He argues that instead of delving further into the sociology of institutional medicine it would be more fruitful to embed broad issues of health and sickness into a wider study of society, including ideologies, power, and occupational structures, and to treat medical issues as part of a discipline-based sociological approach, or what he calls 'medicine-in-sociology' (Turner, 1987: 1). Turner then goes on to show, chapter by chapter, how this might be done, linking in a very convincing manner the study of medicine into wider social structure through the discipline of sociology. Turner thereby 'reclaims' medical sociology as a part of sociology, and ties it into the central conceptual and theoretical issues of social theory.

What Turner achieves for medical sociology is one solution to the atheoretical nature of housing studies. Reconceptualising housing studies as part of separate disciplines, such as sociology, can, and I would argue should, be a first step in decomposing housing studies prior to rebuilding it along truly interdisciplinary lines. In the chapters that follow I intend to begin the work of just such a reconceptualisation in terms of sociology.

EMPIRICISM AND THE PROBLEM OF EPISTEMIC DRIFT

The problem that Turner (1987) grappled with concerning medical sociology is similar to that of defining the core of housing studies – despite the fact that the former is an intradisciplinary field while the latter is multidisciplinary – in that in both cases there is a clear tendency for the subject matter to descend to the lowest common denominator. In medical sociology Turner was concerned to raise the focus from the narrow one of medical institutions to broader issues of health and sickness. In housing studies the descent to the lowest common denominator of bricks and mortar is a comparable problem.

This phenomenon of the downward drift of subject focus can be described in terms of a concept borrowed from another context, namely that of 'epistemic drift' (Elzinga *et al.*, 1985). This was coined to describe the way in which discipline-based concerns became subtly redefined by an almost imperceptible exertion of long-standing influence upon academics by policy-makers. Transposed to the issue of the descent to lowest common denominators in medical sociology and housing studies, epistemic drift can be seen as the process of de-conceptualisation that takes place in respect of conceptual frameworks applied to concrete social phenomena when researchers are under the pressure of policy-making concerns. The result is that the focus of the field shifts from issues embedded in wider problems within the social sciences to very specific ones defined by the shifting and unstable current policy concerns of administrators and politicians.

Epistemic drift can, and often does, take place in several directions simultaneously. Epistemic drift towards policy concerns and towards lowest common interdisciplinary denominators are characteristic tendencies in housing research. Other drift-like tendencies may take place from time to time, for example towards abstracted empiricism or grand theory, towards micro or macro biases, or towards quantitative or qualitative research methods. An important task of housing researchers is therefore to be alert to the tendency towards epistemic drift in housing studies. This requires a degree of reflexivity and theoretical awareness among researchers, which is encouraged by an awareness of the relationship of housing studies to other areas, as discussed in this chapter, and an awareness of the epistemological grounds of the field, an aspect of metatheory that is the subject of Chapter 2.

CONCLUSIONS

I have argued that the place of housing studies within the social sciences needs to be clarified and developed in a number of respects. The rapid expansion of the field in recent decades, its growing complexity and diversity, and above all the growing interest in theoretical and conceptual issues, make it increasingly important that housing researchers begin to systematise their work in relation to the social sciences. A number of tasks present themselves in this context.

The first task is to begin to obtain a clearer idea of precisely what housing studies is about in terms of conceptual and substantive focus. Is it possible to specify the focus of housing studies in terms of a social dimension, such as the home, or residence? This may well prove to be an intractable problem, but it is worth addressing in order to increase awareness of the issues. In addition, it would be wrong to foreclose the discussion because of the difficulties involved, as even the most pessimistic predictions can sometimes be confounded.

Housing issues need to be more closely related to debates in the disciplines than is still the case today, in spite of progress in recent years. Housing research is too isolated and has become too narrow in focus, with housing issues too often abstracted out of the wider social context. Housing research needs both to draw more extensively from debates and theories in the social sciences and to contribute to such debates with studies of housing. These issues are illustrated in more detail in chapters 3, 4 and, to a lesser extent, 5.

In the long run it would be greatly enriching if housing research were to become interdisciplinary, drawing explicitly on theories, concepts and debates within more than one discipline and applying these to housing in an integrative manner. This would also involve feeding back into the disciplines with findings and applications of concepts and theories taken or adapted from the disciplines. In this way, housing research could make important contributions to broader debates which may have little or nothing to do with housing directly.

Housing research can also draw from, and contribute to, debates and issues in cognate fields. Research on social welfare and the welfare state is one obvious area. Theories of the welfare state, drawn from sociology and political science in particular, form the obvious disciplinary sources of its theorising, while social welfare research provides the cognate field within which housing can be understood as being one element.

Awareness of the relationship between housing studies on the one hand and the social science disciplines and cognate fields on the other enables housing researchers to monitor continually developments in their field. This creates opportunities for 'contextualised reflexivity' in which developments in housing studies can be kept in step with those in other areas. Sometimes it may enable housing research to result directly in the initiation of new developments in other areas.

The importance of housing studies developing in relation to other social science areas cannot be over-emphasised. It provides the stimulus for outside influence and acts as a mirror to the state of the field. In the next chapter I address some of the basic issues around theorising in relation to housing. In doing so, I draw heavily on developments and contributions in other social science areas, and in particular in my own discipline of sociology.

2 The epistemological grounds of housing studies

INTRODUCTION

Although the undertheorised nature of housing studies is well known, little has been done to understand why this is so and what may be done to rectify it. One of the reasons for the low level of theoretical awareness within housing studies is that very little reflexivity is apparent in the work of housing researchers. As a result, the question of how housing studies may be conceptualised and the limits within which it may be theorised tends not to be addressed.

This situation is beginning to change, particularly since the mid-1970s, and there is now a growing awareness of the theoretical issues that lie behind housing questions and that can help to inform their analysis. This growing theoretical awareness has been partly in response to the influence of Marxist perspectives. Thus, for example, during the 1970s the nature of the state and its consequences for the nature of the housing market received some attention (Clarke and Ginsburg, 1975; Forrest and Lloyd, 1978). More recently, there has been interest in shifting the emphasis from narrow policy issues to broader social ones surrounding the nature of housing provision (Ball, 1983; Ball et al., 1988).

But even outside of Marxist theory there has been growing theoretical interest since the late 1960s. The most important development has been the pioneering work of Rex and Moore (1967) in which the concept of housing classes was developed to understand inner city ethnic disadvantage. This led to a substantial literature on housing classes, the basic argument being that housing advantage is a major determinant of social inequality (Bell, 1977; Couper and Brindley, 1975; Rex, 1973; Saunders, 1978). Other developments have been work on urban managerialism (Pahl, 1969), theories of power (such as élitism and pluralism) in housing (Dunleavy, 1981), the central and local state (Dickens et al., 1985; Saunders, 1986), and the concept of structuration developed and applied in a number of contexts

(Sarre, 1986; Saunders and Williams, 1988). Even methodological issues are emerging in comparative housing studies (Harloe and Martens, 1984; Dickens *et al.*, 1985).

This burgeoning interest in conceptual and theoretical issues is promising and potentially of great importance for the further development of housing studies. But much remains to be done. Coverage is still patchy and few debates are followed through or linked together in broader themes. Two of these themes – the role of the state as conceptualised in housing research and theorising social change in comparative housing research – are taken up in Chapters 3 and 4 respectively. In this chapter I take up and develop more fundamental prior questions concerning the grounds of knowledge of housing studies: questions which have rarely, if ever, been addressed by housing researchers. If a more theoretically aware housing studies is to be encouraged, then more is required than merely pointing out theoretical lacunae in existing housing research. Addressing root questions concerning the basis upon which housing studies rests lays solid foundations for further progress.

The elaboration within housing studies of the theory of the grounds of knowledge – or epistemology – is therefore an important prerequisite to developing a theoretically informed housing studies. In this chapter I discuss some of the epistemological issues which must be addressed if we are to understand what housing studies is and how it can best be developed (see also Kemeny, 1987a; 1988). In doing so I explore some of the broader, taken-for-granted issues that lie behind the field of housing studies.

I divide the discussion into a number of different elements, taking a 'building blocks' approach: beginning with the constituent elements out of which housing studies is comprised and building up to the larger issues. I start by questioning some of the fundamental concepts that housing researchers tend to use unreflexively: concepts such as 'household' and 'dwelling'. I then consider questions underlying methodology, particularly concerning the social construction of official statistics, which normally constitute the empirical basis for first-order concepts. This leads naturally to a consideration of the intimate link between the generation of 'data' and concepts that are derived from these. Following on from this, I examine the way in which higher level composite concepts such as 'overcrowding', 'dwelling type', 'housing subsidies' and 'forms of tenure' are constructed in an intimate interweaving between data, methods, and concepts which constitutes the subject of debates and controversies in housing research. At a still higher level of generalisation I then consider the questions of how housing problems are defined, and why at certain times some issues but not others become housing problems. I argue that housing researchers have too easily accepted the way in which policy-makers define what is, or is not, a housing problem, and that housing researchers need to begin to take a more active part in defining housing problems, informed by wider theories and debates.

My argument in this chapter boils down to two major points. The first is that housing researchers need to be much more reflexive in their analysis. By this I mean that housing researchers need to think more about what they are doing and why, rather than simply doing it. The second and related point is that a more critical approach is needed, in the sense of questioning and challenging taken-for-granted positions, in order to be able to move towards a situation in which housing researchers rather than policy interests increasingly define the research agenda. My position here closely resembles that of Wright Mills, as elaborated in *The Sociological Imagination* (Mills, 1959) especially in terms of reflexivity, questioning taken-for-granted positions, his focus on middle-range theories, his tying in of individual and societal levels in terms of personal troubles of milieu and public issues of social structure, and above all his critical stance.

However, it would be too ambitious to claim that this discussion does any more than introduce epistemological issues and sketch out in a very preliminary manner some of the possible ways forward that might be taken in developing an epistemology of housing studies. The discussion below must therefore be seen as catalytic: a means of opening up broader questions and suggesting items for a research agenda that housing researchers need to address.

BASIC CONCEPTS

Every discipline or field has a set of basic concepts that underlie analysis: for example in political science 'the state', and in sociology 'social structure'. Sometimes these concepts are subjected to analysis in order to attempt to arrive at a satisfactory definition. But normally they are invested with an assumed certainty that we know what they mean, and second-order concepts are derived from these in order to build up theories and explanations of specific phenomena (for example, types of state, – totalitarian or liberal–democratic, and under which sets of circumstances each is most likely to be found).

Housing studies has its own taken-for-granted 'first-order' concepts that underlie virtually all housing research. Two examples are those of 'household' and 'dwelling'. At first glance, it might be thought that these are quite unproblematic, and that we all 'know' what a household is, and what a dwelling is. In fact, as I have indicated in Chapter 1, there is considerable ambiguity surrounding the meanings of these basic terms.

This ambiguity is reflected in the vague and confusing definition given in the instructions to census enumerators:

> 29. *A household* is either one person living alone or a group of people (who may or may not be related) living, or staying temporarily, at the same address with common housekeeping. Enumerators were told to treat a group of people as a household if there was any regular arrangement to share at least one meal a day, breakfast counting as a meal, or if the occupants shared a common living or sitting room.
>
> (Office of Population Censuses and Surveys, 1981: 6;
> emphasis in original)

This definition is vague in the extreme. What counts as 'an address', or 'common housekeeping'? What counts as a 'regular arrangement' as against an irregular one? What counts as a 'common' living or sitting room? Although the definition was qualified by a page of examples for guidance, this failed to remove much of its imprecision and vagueness, amounting to an attempt to circumscribe the definition in terms of the length of time and the proportion of time that people ate and slept in a household. Thus, aux pairs were defined as part of the household only if contracted for at least a year (but not if for a shorter period); boarders if they returned to another address at weekends (but presumably not if they returned to another address on other days); children at boarding school, university, etc. (but not if married, and not if working abroad for however short a time); persons in an institution only if they had been there for 6 months or more (though the enumerators were instructed to accept the head of household's decision regarding classification); spouses who worked away from home, abroad, or in the forces (but not if separated and only visiting 'occasionally').

These cases give an insight into the extreme complexity of the everyday definition of 'household' that housing research is grounded on. It is a measure of the complexity of this issue that individuals are themselves often unclear about which household they belong to. This may be more widespread than is first thought. It could include many young people who live in digs but often stay at the parental home; separated couples in transition from marriage to divorce; children who are the divided responsibility of divorced parents and live in each home half of the time; and highly mobile people with more than one home. The possibilities are as limitless as human variety and ingenuity.

Over and above such inherent ambiguities are those ambiguities that people consciously create in response to such things as tax liability or subsidy eligibility. We therefore need to be aware of the sources of deliberate manipulation of responses to interview questions concerning household membership. These may be appreciable. For example, among social security recipients the cohabitation rule may distort response. With the introduction of the poll tax, we might expect this sort of distortion to increase markedly, since poll tax liability is based directly on the ambiguous tension between the concepts of household and dwelling (most clearly evidenced in the fact that, despite its being a tax on people, empty dwellings are taxed at a penalty rate). This might easily lead to widespread registering of one member of the household at the holiday home in order to reduce poll tax liability. A similar use in Sweden of the holiday home (owned by 25 per cent of households) is evident, where the priority given to single parents in access to daycare facilities leads to the false registering of one of the parents in the holiday home on a scale that can only be guessed at. Such distortions will often be used to play upon the natural ambiguity that many ordinary people may feel in terms of their household membership.

The concept of 'dwelling' is equally problematic as that of 'household'. It will have been noted that it was not possible to avoid bringing aspects of the dwelling into the discussion of the ambiguities surrounding the definition of a household. Households in the census were defined partly in terms of the dwelling they occupied: notably by living at the same address and sharing dwelling facilities. But beyond this, it is difficult to know at what point a building can be said to be divided into separate dwellings. That is, the definition of a dwelling is itself highly complex.

Thus, for example, some forms of collective housing or institutional homes may or may not be defined as separate dwellings depending on the number of separate facilities they possess. Old people's housing is a case in point. Some forms of sheltered housing may pass muster as separate dwellings, but may not if some facilities are collectively organised. Holiday homes incorporate different sources of ambiguity. For how much of the year can one live in a building before it is classed as permanently dwelled in? How 'permanent' a structure does it have to be before it is classed as a dwelling, and what constitutes 'permanence'? Many dwellings are physically joined to one another, and in some cases access to one cannot be made without going through another. Are these separate dwellings or not? And what about houses which include a part with its own street access (such as a granny flat) but with direct door access to the house? Is the part a separate dwelling or not?

Part of the problem is that, just as a household is partly defined in terms of a dwelling, the reverse is also the case: an edifice is defined as a dwelling partly in terms of the definition of a household. This means that the ambiguities of the concept of a dwelling are compounded by the ambiguities inherent in the definition of a household. Thus, for example, a granny flat is more likely to be classed as a separate dwelling if someone lives in it more or less regularly than if it is lived in more or less irregularly!

The ambiguities are compounded when basic concepts are then used to build second-order concepts such as 'overcrowding', by combining the concepts of household and dwelling in various ways, or 'forms of tenure', by combining the concepts of dwelling and household ownership status. Of course, there will always be ambiguity in concepts, as Garfinkel (1967) has clearly shown. But for research purposes it must surely be fundamental, first, to examine concepts with great care and, second, to reduce ambiguities and sources of dissonance to a minimum in their operationalisation.

A major task in developing a theoretically grounded housing studies must therefore be to clarify concepts and air problems surrounding fundamental definitions. It would, for example, be useful to have a rough idea of the extent to which ambiguities in the definition of a household are reflected in the unreliability of household data. There is a strong case to be made out for major, empirically based conceptual work to be done on estimating the degree to which household data are reliable, and to obtain at least a very rough approximation of the extent of possible error in them (for a preliminary attempt at this see Todd and Griffiths, 1986). The concept of dwelling is probably less problematic than that of household, but could also be usefully analysed to produce rough estimates of the likely degree of variation in numbers created by major ambiguities. Finally, work can usefully be done on disentangling from one another, to some extent at least, the concepts of dwelling and household.

In the final analysis, however, the concept of household is likely to prove to be too much of a hybrid and too intimately bound up with the concept of dwelling to constitute a useful starting point for the study of housing, as the very word 'house-hold' clearly suggests. There appear to be two alternatives. One is for the concept of the individual to replace it. Many households in any case comprise only one person and, since members of the same household will often have different – and sometimes conflicting – interests, it clearly makes more sense in certain circumstances to take the individual rather than the household as the basic starting point for housing research. However, since many dwellings are occupied by more than one person, the act of sharing, or 'holding', a house in common makes the concept of household indispensable. An alternative to the individual as the base concept for housing research, for the purpose of solving many problems at least, may be to treat households in dwellings as one unit of analysis and to talk of 'households-in-dwellings'. This is a usage that I intend to employ in much of the subsequent discussion in this book.

In order for housing research to be carried out at all, the problematic nature of fundamental concepts is conventionally ignored. This is inevitable and necessary, at least to a certain extent, if housing research is not to be abandoned as impossible. But, at the very least, housing researchers need to be sensitised to the problems of the data that they take for granted as underlying their work. If this leads to greater scepticism and a more critical approach to the quality of data, then housing researchers will have made a start on addressing fundamental methodological issues.

METHODOLOGICAL ISSUES

There is a large literature in the social sciences on the grounds of methodology. Much of this derives from the debates over the relative merits of qualitative and quantitative research (Denzin, 1970). Qualitative research has for long been treated as a poor cousin in sociology, in its attempt to gain recognition as a 'science', the hallmark of which was seen as the use of statistics and other quantitative measures of social phenomena.

The reaction of qualitative researchers was to counter-critique positivistic methodology in one of two ways. The first was by showing how the grounds of quantitative methods are in fact built on the use of categories which researchers derive from implicit theory and which they impose upon the data. Cicourel (1964), for example, showed how much research is based on such deductive categorisations, which necessarily involve a degree of arbitrariness in relation to the data in a manner which Warren Torgeson had termed 'measurement by fiat' (Cicourel, 1964: 12). This provided qualitative methods with a theoretical basis for their use, in terms of avoiding distortions that may arise when external categories are imposed on data in such a way as to 'do violence' to them, and in terms of 'taking the actor's definition', leading to a 'naturalistic' method grounded in empirical data (Glaser and Strauss, 1967). Taking the actor's definition became the hallmark of sensitivity and accuracy of qualitatively oriented researchers in describing and analysing social situations.

The second was to show how what was being measured was in fact not some objective 'social fact' but the outcome of social processes which could be changed. An early target for criticism was the work of Durkheim on suicide, where he tried to demonstrate the existence of suicide as a 'social fact' (Durkheim, 1952). Douglas (1967) and Atkinson (1978) showed, for example, how Durkheim's study of suicide rates, based on official suicide statistics, told us more about how these statistics were collected than about the frequency of suicide: for example, how the religious stigma attached to suicide in Catholic countries led to massive under-reporting, and how, when the social organisation of the collection of suicide statistics was altered – from being the responsibility of priests to that of coroners – the suicide rate suddenly increased.

But criticism extended to the use of all official statistics by social scientists, and this became the subject of a lively debate (Hindess, 1973; Kitsuse and Cicourel, 1963). Especially singled out for attention were statistics used in criminology: on criminal charges, arrests, convictions, and other fundamental indices of crime rates. It was argued, for example, that arrest and charge statistics tell us more about the social organisation of policing than the frequency of crime, and that changes in the organisation of policing – for example from foot-patrolling to car-patrolling, or a policy innovation, such as linking promotions to successful prosecutions – led to changes in the types and number of arrests that were made (Black, 1970; Ditton, 1979). It was also argued that media reporting can generate an increase in the production of crime statistics in the form of a self-fulfilling prophecy, by sensitising police and courts to specific issues (Fishman, 1978). Such heightened attention could, under certain circumstances, create the phenomenon of imitation, thereby creating an escalating cycle of action and reaction leading to a 'moral panic', the classic example being the mods and rockers phenomenon of the mid-1960s (Cohen, 1970).

These kinds of issues are rarely, if ever, discussed in housing studies, despite the wide range of qualitative and quantitative housing research that has been carried out. The most developed methodological debates are, as indicated earlier, those around 'realism'. Beyond this, what is needed first and foremost is a reflexive awareness of the main thrust of housing research methodology.

Given the extensive use of official statistics by housing researchers, analysis of the manner in which these are socially constructed and the social organisation around their generation would be a fruitful starting point. This would involve a study of the social organisation of the production of housing statistics, and of the kinds of built-in biases that such social organisation tends to generate. The point would be to show the extent to which housing statistics are a product of the organisation that generates them rather than a gauge of the phenomena of which they appear to be measures.

This could be done, for example, with building society lending statistics, by examining the ways in which the social organisation of building society branch lending produces loan statistics that reflect local branch policies in combination with the guidelines of the central office. It would then be possible to see the extent to which changes in branch managers and other personnel, and changes in procedures for allocating loans impacted on lending practices and hence on the statistics that were generated. It might also be possible to obtain some idea of the way in which lending practice is influenced by media and political attention to produce a parallel phenomenon in housing to the 'crime wave' effect, sometimes leading to a minor 'moral panic' at the perceived threat to owner occupation, such as that inducing British government intervention to boost mortgage lending in 1974.

The point of such work would be not only to research areas of intrinsic interest in the field but also to use such research to lay bare basic methodological problems in the use of official statistics and survey data of all kinds. This would raise awareness of issues and problems which at present are almost completely neglected, as well as opening up new areas of research.

THE RHETORIC OF CONCEPT-FORMATION

We often think of research as something objective; as a process of the unearthing of facts, to reveal some essential truth about the subject matter. The very term 'unearthing' suggests that these facts were there all the time and were only waiting to be 'dug up'. This is often how researchers themselves portray what they are doing. Another way of looking at this is that research is about persuasion and attempting to convince others, and particularly other researchers and policy-makers, to accept one set of understandings and interpretations on particular issues. In this view, a body of data is constructed or assembled for research purposes to use as evidence in an argument. This is what Wright Mills described as the 'craft' of research (Mills, 1959: appendix). There is therefore an intimate relationship between data, methodology, and concepts. Data are not of themselves meaningful until they are interpreted by the researcher and a meaning is imputed to them. They are not free-floating facts, independent of the researcher, which are tapped into, as it were. Rather, they are socially constructed in the act of conducting research, and this is over and above the social construction of the data themselves, as in the case of official statistics, by whichever agencies are responsible for their compilation. Research, in this view, is an essentially rhetorical activity involving the constructing of arguments in order to introduce new concepts or to change existing ones as part of a campaign of persuasion. A useful analysis of natural science as the art of persuasion can be found in Latour (1987).

Much work has been done in the sociology of science on the research act. Here I want to draw on the work of Latour and Woolgar (1979) in their study of endocrinological research carried out at the Salk Institute using ethnographic analysis (for a fuller treatment of the application of this to housing research see Kemeny, 1984). Latour and Woolgar show how scientific facts in the endocrinology laboratory – the data derived from experiments – cannot be understood apart from the technology, or 'inscription devices' to use Latour and Woolgar's term, used to produce them. The equipment used to abstract and process data, such as pieces of mice, include mixing machines, computers, and various items of complex equipment which are designed to convert raw material into data by means of a highly complicated multi-step process. These data in turn comprise the ammunition used to back an argument in a process of formulating and justifying concepts.

Latour and Woolgar argue that the research act consists of manipulating data in such a way as to support an assertion, or claim, that certain conditions, or objects, have the status of being real rather than merely hypothesised. As they put it, this involves converting 'artifacts' into 'facts'. This process of transformation involves what the authors describe as 'splitting'. Taking the example of TRF (thyrotropin releasing factor), they show how its status changed from that of a hypothesis, or artifact, into a fact established through experimentation. Somewhere along this process, the status of TRF began to change from artifact to fact and possessed both qualities in various degrees (Latour and Woolgar, 1979: 179–80).

Ultimately, splitting will result in stabilisation, either with the facticity denied or affirmed, depending in part on how good an argument can be made with the data which have been produced. The end result is that what the authors term an 'inversion' takes place, in which the hypothesis becomes redefined as a fact which has been there all the time, waiting to be 'discovered'. Thereby, the fact is held to have preceded the statements about it.

The conversion of artifact into fact through splitting and the ultimate inversion of the order of events can be demonstrated in housing research in a number of ways. As we have seen, basic concepts, such as 'household' and 'dwelling', tend to be both stable and largely unchallenged. It is second-order concepts that are more liable to instability. A particularly clear example, and one involving both housing researchers and policy-makers, is that of the concept of 'overcrowding' in Swedish research and housing policy.

Swedish housing is built according to clearly defined standards, one of which is the level of occupancy. To ensure that dellings do not exceed this, an overcrowding norm is set. The level is in part determined by what housing administrators and politicians believe to be acceptable minimum standards but naturally there is potential disagreement about where the line should be drawn.

As part of the debate to raise standards, Eriksson and Lindqvist (1983) attempt to destabilise the current definition of overcrowding and to obtain acceptance of a new, higher norm. To do this, they marshal a whole range of data, including computerised census statistics and architectural layout diagrams. Their case hinges on arguing that official statistics on overcrowding underestimate overcrowding because of the manner in which they are constructed (for example, by treating all rooms as approximately equal in size, by assuming all couples will share a room, and by failing to take into account the special space needs of certain kinds of household such as remarriage families).

The strategy is therefore to argue for a higher overcrowding norm by pointing out the deficiencies in official statistics but not to the extent of calling their basic validity into question. This requires a delicate balancing act, carefully tailoring the argument to produce an elegantly formulated and convincing case.

The same logic applies to the concept of household that is so fundamental to the question of overcrowding. Problems are pointed out in the way in which households are defined but only to the extent of supporting the case being made. Thus, for example, the researchers' argument that certain types of household, which are quite significant in number, have special space needs because of the special structure of the family (such as the 'mixed' remarriages mentioned above, or multi-generational and extended kin immigrant households) touches upon the frailty of the concept of a household and how it is defined, but without launching a fundamental critique of it.

Latour and Woolgar take their argument further than all this in one respect. They argue that there is no simple one-to-one correlation between speculation and artifact status on the one hand and unquestioning acceptance and fact status on the other. Rather, we are concerned with a cumulative body of statements backed up by arguments and data which over time gradually shift status from artifact to fact or vice versa.

Specifying this in more detail, Latour and Woolgar break down the status of concepts into a continuum comprising a five fold classification, ranging from taken-for-granted status to speculation (Latour and Woolgar, 1979: 76–7.) They argue that the 'art of science' is to move statements along the continuum from speculation to taken-for-granted status in order to convert artefacts into facts. Publications (buttressed by complex cross-referencing that Latour (1987) describes in terms of calling witnesses to support a case), and complex laboratories comprise the resources needed to establish a convincing case, and to build up a following among other researchers that can ultimately become a school of thought aimed at turning a hypothesis into a fact.

Thus, in the example of overcrowding, we can see how a gradual shift in the perceptions of researchers and policy-makers can come about, leading either to acceptance of the new definition or to the formation of two or more schools of thought on the issue. The success or otherwise of researchers in moving statements about overcrowding along the continuum towards taken-for-granted status depends not only upon the existence of a number of committed researchers able and willing to pursue the debate but also upon the speed with which housing standards rise or fall during the period of the debate. Clearly, during periods of rising housing standards it will be easier for the proponents of the new definition to change the status of the concept in the desired direction, while at other times it will be harder.

The concept of 'housing subsidy', in terms of whether tax deductions should be classed as a subsidy or not, is a controversial issue for different reasons. Here the issue behind the definition of the concept is whether owner occupiers are being subsidised because they obtain mortgage interest relief, in contrast to renters who do not. This issue may be more difficult to change because of the deeply entrenched ideological issues involved, and the fact that as yet no-one has succeeded in marshalling the requisite evidence and arguments to move statements about housing subsidies decisively towards fact status decisively enough.

In addition to well established concepts that may be challenged, new concepts continually arise as researchers experiment with new explanations and approaches. The concept of 'housing classes' (Rex and Moore, 1967) is one such which has been introduced and continues to have some usefulness. That of 'structuration' (Sarre, 1986) is more recent and it remains to be seen whether it will prove useful or will eventually drop out of usage.

WHAT IS A HOUSING PROBLEM?

The rhetoric of concept-formulation leads into broader questions of how housing problems are defined. If housing concepts can be understood in terms of an advocacy model, does this not have implications for the very manner in which larger housing issues are understood? The actual way in which we frame questions in this view becomes problematic.

Concepts such as 'household' form the basic building blocks of the ways in which we understand the social world in which housing issues are located. But the perspectives taken on these issues in traditional policy research are equally questionable. Hitherto, they have tended to be received and largely unquestioned. They tend to be those issues and problems which are defined as such by politicians, housing administrators and housing market interests. We need to begin to understand the ways in which 'housing problems' are defined by interest groups and policy-makers and to develop an awareness that the perspectives we apply and the housing problems we define need not be merely 'received' from vested interests but can, and should, be consciously worked through and chosen by housing researchers.

There is a large literature on the definition of social problems which argues that what is or is not defined as a social problem is the result of the ability of particular interests and social groups to impose their own definitions upon the ways in which issues are conceived (Cohen and Young, 1973; Green, 1975; Hilgartner and Bosk, 1988; Kitsuse and Schneider, 1982; Kitsuse and Spector, 1973; Mankoff, 1972; Safa and Levitas, 1975; Spector and Kitsuse, 1973, 1977). An early example of such work was Gusfield (1963), whose study of the American temperance movement showed how temperance societies struggled with varying degrees of success over the decades, swinging from prohibition to more liberal values followed by reaction, to have alcohol use defined as a 'problem'. Since Gusfield's book, there have been studies of almost everything from margarine (Ball and Lilly, 1982) to nuclear power (Useem and Zald, 1982).

A number of examples can be briefly reviewed to illustrate the relevance of this kind of approach. Governments wishing to reduce housing expenditure, particularly in times of financial retrenchment, may wish to change definitions of 'housing need' as a means of reducing public expectations of housing standards (for example, by encouraging delayed home-making by young adults so that they stay in the family home longer and thereby reduce housing demand).

The redefinition by British local government of the concept of 'homelessness', originally legislatively defined to strengthen people's rights to housing, to exclude all but a tiny number of households and so to circumscribe rights to housing is another important example of defining a problem. The impact of media attention – notably in the film *Cathy come home* in the 1960s – on heightening the awareness of homelessness and thereby putting pressure on governments to take action is only a particularly dramatic example of the way in which it is possible to redefine housing problems. This would, in fact, make a fascinating historical study of power and conflict over the social construction of a housing problem, tracing the debate through the media and its effects on politicians and the drawing up and passing of homelessness legislation, culminating in the undermining of the legislation, through the redefinition of homelessness in the courts by local authorities. It is indicative of the lack of independence of housing research from policy-making that such a book has not been written.

Mortgage arrears is another area where ideological meanings may be important. Governments committed to sustaining high rates of owner occupation are likely to play down the significance of the numbers of defaulters. By contrast, those who want to highlight the problematic nature of owner occupation, perhaps as part of a campaign to increase the amount of public rental housing, are likely to emphasise this as a 'problem', particularly during times of high interest rates or static and falling house prices.

Housing researchers therefore need to begin to question the very manner in which definitions of housing problems are sustained or changed and the impact that this has on policy. The large literature on social problems can and should be drawn on to provide a starting point. The point of this is not merely 'institutionalised contrariness' but rather to understand the influence of policy-makers on housing research and to challenge taken-for-granted definitions of what constitutes legitimate and urgently needed housing research. This in turn is a first step towards the conscious control of the research agenda by researchers rather than by policy-makers. It constitutes a politicisation of the research agenda, which demands of housing researchers that they acknowledge, and take a stand on, the inevitablity of their involvement in the social construction of housing issues.

CONCLUSIONS

The wide-ranging and necessarily brief treatments of a number of major issues in this chapter provide the starting point for formulating the grounds of theory for housing studies, and we are now in a position to bring the discussion together. This may be summarised as comprising a number of elements.

One task is to examine the key concepts that underlie housing research: how they are constructed, their weaknesses and limitations. This is not only a worthwhile exercise in its own right but also provides some crude measure of the extent to which data generated by their use are reliable in different empirical contexts.

Second, work is needed on the grounds of methods to raise awareness of methodological problems. In particular, we need to begin to evaluate critically the reliability of official statistics, as has been done in criminology, media studies, and the study of social problems.

Third, there needs to be greater awareness of the rhetorical nature of housing research, and the way in which housing debates interweave concepts, data, and methods in the context of specific policy debates.

A fourth task is for housing researchers no longer to accept unquestioningly 'received' definitions of what does and does not constitute a housing problem, and to begin to reformulate problems and to develop problems of their own, closely tied to theoretically informed issues. researchers need greater awareness of the way in which housing problems are defined, rather than accepting definitions in the wider society as given.

The epistemological grounds of housing studies therefore provide a wide range of new issues for housing researchers to address. Explicating these will provide a much needed boost to theoretical work in the field, and greatly increase the reflexivity of research.

Part II
Bringing theory back in

3 Returning to the state in housing research

INTRODUCTION

There must surely be unanimity among housing researchers that the state plays a major – some might say dominant – role in the housing market. No study of housing can be carried out without a considerable part of the discussion being devoted to an analysis of the state and its direct and indirect influence on the provision and consumption of housing. Yet the extent and nature of state intervention in housing has never become established as a major theoretical and conceptual issue. Traditional policy analysis did not address this or indeed most wider conceptual questions. Even in more recent housing research, theorising around the role of the state remains minimal and unsatisfactory.

But if theorising has been neglected, one might have expected that at least the debates in sociology and political science over the role of the state would be followed with keen interest and that this would be reflected in the housing literature. Nothing could be further from the truth. Since the autonomy of the state became a major issue over a decade ago there has been little evidence of housing researchers' awareness of, let alone interest in, the debates that it has engendered.

Nor is this merely some aberrative lacuna in the housing literature. Since the degree of autonomy of the state is relative only to other sources of power in society, it cannot be understood outside of the wider debates over identifying the locus of power in modern societies. Yet with a few major exceptions – most notably Dunleavy and Saunders – there is almost no awareness shown in housing studies of the vast literature on theories of social power, stretching back to the long classical tradition of élitist theory from Plato through Machiavelli to Mosca and Pareto and forward to the large early postwar literature on élitism and pluralism and the more recent Marxist, neo-pluralist and corporatist theories.

It could perhaps be argued that the strong policy emphasis of housing studies means that broad-brush macrosocial and political theories of power are not seen as relevant, and that this is why there has been so little interest in them. So what of middle-range theories of power which would be more directly relevant to empirical research? In fact, there is little or no interest shown by housing researchers in some of the more detailed debates within the power literature. Thus, the work of Bachrach and Baratz (1962, 1963) on 'hidden agendas' and of Crenson (1971) on 'non-decisions' are prominent and widely debated examples of the kind of middle-range theories based on specific power studies that are directly applicable to, and testable in terms of, housing policy. Yet there does not appear to be any housing research which uses them. It is therefore necessary to conclude that housing researchers have felt that there is little, if anything, to learn from theories of power.

This chapter begins by contextualising the issue of the autonomy of the state in the changing conceptions of power that followed in the wake of the development of postwar political sociology, before summarising the recent debates in political science over the autonomy of the state as a social institution. The housing literature is then reviewed to show how, partly as a result of growing interest in social theory in recent years, an essentially 'non-statist' perspective on power has been in the ascendant, though in largely unexplicated form. Finally, I argue for an explicated approach to the role of the state in housing that integrates elements of both statist and non-statist perspectives and that brings the issue more to the forefront of housing analysis.

THE RETURN TO THE STATE DEBATE IN POLITICAL SCIENCE

Until the 1950s, political science tended to focus disproportionately on political processes narrowly defined. The study of political constitutions, legislative and cabinet procedures, political leadership, decision-making, and psephology gave political science an almost psychologistic bias. Theorising tended to be limited to either philosophical analysis of the classical literature on power and rights (Plato, Machiavelli, Paine, Locke, Rousseau, etc.) or constitutional philosophy (e.g. Bagehot, de Tocqueville). The development of modern postwar social sciences therefore had a profound effect upon political science. Along a broad front of research, from community studies (Dahl, 1962; Floyd Hunter, 1953; Vidich and Bensman, 1958) through studies of national and regional élites (Aaronovitch, 1961; Mills, 1956; Lipset, 1959), the emphasis shifted away from politics and towards the study of the social basis of power in terms of class, organisations (such as the military), and other vested interests, out of which grew pluralist and élitist theories of power.

One of the consequences of this direction of attention towards broader social processes was a tendency for the early postwar political science approach to power in narrow political terms to be seen as hopelessly blinkered and limited. In the general hubris over the inadequacy of 'psephology' and 'great man' explanations of power, there was widespread condemnation and derision of the study of politics which was seen as 'political reductionism'. Instead, it was argued that the key to understanding power was to be found by paying attention to its social bases, and, in the plethora of societal power studies that followed, political institutions took second place as a focus of analysis. It became axiomatic that state institutions lack real power and are by and large vehicles for the expression of vested interests, and that political institutions were just one of a number of arenas in which power was exercised.

The rise of Marxist social science in the 1960s served, if anything, to reinforce this tendency. Marx's dictum that 'the state is the executive committee of the ruling class' strikingly epitomises this position, and the economic class approach of Marxism is a major reason why broader social perspectives on power are dominant within Marxist analysis. Thus, even though some autonomy is granted to the state (Offe and Ronge, 1975) and even though there is debate over the manner in which class interests are mediated by the state (see Miliband, 1969; Poulantzas, 1973), the strength of Marxist perspectives on power lies in the analysis of classes and their struggles: first and foremost in their economic struggles and how these then become mediated and refracted through political institutions.

But, as so often happens, a reaction against the neglect of the state eventually set in, though it took many years to manifest. There has been, since the late 1970s, an attempt to return to the examination of the state as a source of power in its own right. Turning the criticism of political reductionism against its critics, the reaction took the form of a counter-critique of what was seen as 'societal reductionism' which ignored the patently powerful state apparatus with its armed forces, police, local agencies, and taxation system. By the early 1980s, it became increasingly clear that treating the state as a passive black box through which social power is exercised was woefully inadequate and grossly distorting. 'Bringing the state back in' as a central problematic in political and social analysis therefore took the form of an attempt to rewrite the research agenda (Evans *et al.*, 1985; Nordlinger, 1981, 1987; Skocpol, 1979), with the Committee on States and Social Structures of the US Social Science Research Council taking an active part.

Skocpol (1979: 24–33), one of the prime movers behind the return to the state in political science, identifies two major tasks of the state as a social institution: to maintain order and to compete with other states. Maintaining order involves the sustaining of an administrative structure, including taxation and the maintenance of social order, and where necessary making concessions so as to forestall social unrest. Skocpol argues that the state may implement the interests of dominant classes but that in the final analysis it will often support subordinate classes so as to preserve internal order. However, the prime role of the state is to protect the domestic order from international competition and aggression, and to extract maximum benefits from other states through competition, war, or colonialism. These two roles – the preservation of internal order and the defending and fostering of domestic society against external opposition constitute unique characteristics of the state, shared with no other social institution, that have major consequences for the organisation of society.

The recent interest in theorising the state in political science has stimulated debate in sociology, where interest has been particularly focused on the territorial and spatial dimension of the state (Collins, 1986; Giddens, 1985; Mann, 1984). Mann argues, for example, that state power must be understood in terms of infrastructural co-ordination within a spatial boundary, these two dimensions varying to produce different forms of state power (Mann, 1984: 191). The 1980s also witnessed a considerable increase in awareness of the importance of the state in Marxist theory (Block, 1987; Clark and Dear, 1984; Held, 1984; Jessop, 1982), particularly in terms of ideological articulation.

This 'return to the state' and its concomitant critique of what is seen as 'societal reductionism' in the dominant structural functionalist, pluralist, élitist, corporatist, Marxist, and other perspectives, has been evaluated and largely rejected by Almond (1988) in a lengthy and detailed examination. He argues that traditional pluralist and structural functionalist perspectives, with which he has been closely identified, did not in fact neglect political institutions. That in turn has led to counter-criticism from the 'statist' perspective in a symposium on the issue published in *American Political Science Review* (Fabbrini, 1988; Lowi, 1988, Nordlinger, 1988).

It is clear that issues concerning the role of the state in modern society are once again on the research agenda, particularly in political science. After many decades during which explanations of power were sought disproportionately in the wider social structure, there is now a growing awareness that perhaps the state itself is not just a passive medium of social power, nor even merely 'refractive', but is a major actor in power struggles in its own right, and one, moreover, with a central role to play. The return to the state movement in the social sciences has provided a much-needed redressing of the balance between the relative importance of civil society and the state. It is here that housing research has much to gain from engagement in these kinds of issues. We are now in a position to consider the unexplicated approach taken to the state in housing research.

THE STATE IN HOUSING RESEARCH

Traditional liberal–interventionist housing policy analysis has much in common with early postwar political science, in its emphasis upon the role of government in the housing market. Housing research can, without too much distortion, be seen as an applied form of early political analysis. Similarly, the emergence of neo-Weberian and neo-Marxist perspectives in housing research in the 1970s mirrors – with a time lag of a generation – the critique of atheoretical political reductionism and the move towards the analysis of the social basis of power. But the special nature of housing studies, in association with the time lag, have combined to give the debates in housing studies a very different emphasis. Ball *et al.* (1988: 19–20), for example, criticise traditional housing research not only for being too state centred but also for having a pluralist bias and a one-sided focus on tenure, or what Ball (1983) has disparagingly termed 'the politics of tenure'. This identification of state-centredness and pluralism with housing studies contrasts strongly with the identification of society-centredness and pluralism with 1960s political sociology.

It is clear that among the more theoretically aware generation of housing researchers that emerged in the 1970s, there was a strong move away from the study of policy and politics. As was the case with the tradition of societal reductionism in political sociology during the 1950s and 1960s, an unexplicated version of the societal reductionist perspective has been in the ascendant in neo-Marxist and neo-Weberian housing studies.

The dominance of the social power perspective in political science during the 1950s and 1960s is thereby mirrored by certain prominent tendencies in housing research during the 1970s and 1980s, though in the context of very different research traditions and their development. In part, this emphasis upon the societal bases of power developed in contradistinction to, and as part of the wider critique of, traditional policy-oriented housing research. It was seen as important that research move away from narrow policy analysis and begin to deal with these wider questions. Because of this reaction, the social power perspective in housing research manifests a much stronger bias against the analysis of politics than had been the case in 1950s political sociology. It is therefore possible to characterise some of this work as 'non-statist'. However, in the process, there was little explicit treatment of the role of the state, and it never became contextualised in the debates that were then taking place in the disciplines and in other fields.

In the discussion that follows, a review of some of the contributions to the housing literature on the role of the state will be presented. The main purpose of this is not to cover comprehensively the literature on housing and the state; an impossible task in this context. Rather, it is to outline the debate and indicate the general lines along which it has been developing. The review is divided into contributions from the Marxist and Weberian perspectives. This division is necessarily somewhat arbitrary, as housing researchers generally draw inspiration from a range of perspectives, but serves crudely to indicate the principal influences upon different researchers.

Neo-Marxist contributions

Some of the earliest Marxist work on housing continued to demonstrate an emphasis upon 'the politics of tenure' (see, for example, Clarke and Ginsburg, 1975), albeit within a clearly Marxist framework. But as time went by a decided shift in emphasis took place. Forrest and Lloyd (1978) summarised different approaches to the state in such a way that made clear that the state was increasingly being seen as an arena within which class conflict was being played out. By the end of the decade, society-centred perpectives were in the ascendant, in which housing policy analysis played a minor part. In general, then, Marxist contributions to housing analysis have been characterised by a gradual shift of emphasis away from policy analysis and state action towards non-statist studies.

Lambert *et al.* (1978), in *Housing Policy and the State*, were among the earlier book-length attempts to grapple with issues of housing and the local state in a direct and explicitly theoretical manner. Their approach was influenced by developments in Marxist theory, as well as the early Weberian concepts of 'housing classes' (Rex and Moore, 1967) and 'urban managerialism' (Pahl, 1977). But, unusually for housing researchers, they were also influenced by the symbolic interactionist perspective from sociology, particularly Berger and Luckmann (1967) and Becker (1967).

Lambert *et al.* (1978) manifest all the signs of early interest in theory among housing researchers in their wide eclecticism and in the ambiguities that can arise when different theoretical perspectives are combined. On the one hand, the authors are critical of pluralist explanations of power and argue that 'one of the classic functions of the state [is] that of managing the interests of the ruling class' (Lambert *et al.*, 1978: 170). On the other hand, they argue that there are considerable variations in the style of urban management exhibited by agents of the state. They 'see the style of urban managerialism as having potent ideological force in shaping and reinforcing the dominant pattern of power, influence and profit' (Lambert *et al.*, 1978: 169). The approach taken by these authors is therefore basically characterised as a societal power perspective but modified to some extent by variations in the manner in which agents of the state manage the interests of the ruling class. But this remains the limits of the authors' direct contribution on the matter.

The debate over urban managerialism that took place during the mid-1970s exemplifies in a particularly clear manner the dilemma facing the supporters of the new urban sociology. The original formulation by Pahl (1969) of what came to be known as the urban managerial thesis argued for attention to be focused on those holding key mediating positions in housing and urban planning: building society managers, local government officers, planners, etc. But Pahl came under heavy and sustained criticism, particularly from Marxists, for giving too much weight to agency and not enough to underlying structural factors (for an example of such a debate see the proceedings of the 1975 Conference on Urban Change and Conflict in Harloe, 1975). Pahl subsequently accepted this correction (Pahl, 1977). The problem facing those espousing the social basis of power approach to housing was that processes within housing institutions, and most especially the central and local state, were by definition secondary and derivative in nature. A strong strand of societal reductionism therefore permeated analysis and inhibited a closer examination of the workings of social institutions and, in particular, decision-making processes in political institutions.

Another Marxist contribution to the housing literature which discusses the role of the state at some length is that of Ball (1983), in the concluding chapter (pp. 336–69), and particularly in a section entitled 'Theories of politics and structures of provision' (pp. 342–6). Here, Ball begins his discussion with a brief overview of the wide range of Marxist perspectives on the role of the state in capitalist society. He rejects most of these as being too mechanistic and denying the wide variations in housing policy between capitalist societies. He argues that state policy must be seen as the outcome of class struggles expressed in the political arena, but that these are not simplistic capital–labour conflicts. Rather, they are complex coalitions and groupings of interests which coalesce around issues concerning the existing structures of housing provision and the viable alternative policies that can be developed and sponsored in order to modify these. Ball therefore does not discuss the role of the state as an institution influencing policy in its own right. His main concern is to argue for an essentially non-statist perspective in which the focus is upon structures of provision and the interest groups which manoeuvre around strategies to change or maintain them within the limitations of the capitalist system.

A third important contribution to the Marxist conceptual literature on the state and housing is that of Dickens *et al.* (1985). This is based in part on a comparison between structures of housing provision in Britain and Sweden, and in part on a series of local British studies of housing provision and the struggles around these. This approach is probably the most clearly non-statist of those reviewed here, explicated in a section of the last chapter entitled 'The state and housing consumption' (Dickens *et al.*, 1985: 232–41). The authors argue quite explicitly that, in their view the state does not possess any power of its own but is merely an institution through which social forces in the wider society are expressed in the form of social relations:

> Our starting point is the view that conceptually more important than state institutions, *which in themselves are empty and without power*, are the social relations of a particular state. These social relations are, in turn, an expression of the balance of social forces at a particular time and in a specific economic, social, political and ideological context.
>
> (Dickens *et al.*, 1985: 234; emphasis added).

Turning to the role of local government, or what they term 'the local state', Dickens *et al.* (1985: 238–41) argue that it functions as an intermediary between the central state on the one hand and the locality, with its wide variations in the social relations of housing consumption, on the other. This may take the form either of fine tuning the implementation of national policy to local conditions or of introducing centrally devised forms of housing consumption that are non-local in character. Either way, the local state may find itself torn between central government policy and the pressures emanating from local social relations in response to local-specific housing conditions.

This view of the central and local state is one in which the central state is the instrument of national social relations, while the local state is the agent of the national state, and to that extent *indirectly* the instrument of national social forces, but is sometimes deflected from carrying out this function by local social relations. So the state is always a reflection of social relations in the wider society, both at the national level and at the local level, where it may either be the agent of national social relations or, over certain issues, be captured by local ones.

Finally and most recently, Ball *et al.* (1988) argue that in comparative research within the liberal–interventionist tradition 'the role of the state in determining patterns of housing provision is exaggerated' (Ball *et al.*, 1988: 8). They argue for much less emphasis upon the role of the state in housing markets, and a greater emphasis upon structures of provision, being 'a dynamic process of interaction between markets and the agencies operating in those markets' (Ball *et al.*, 1988: 24). Their perspective is clearly non-statist in emphasis. But even here, they direct their criticism at the *unexplicated* statist perspectives of traditional policy analysis. They do not attempt to develop an explicit non-statist counter-position, nor even to contextualise this within the political science debates.

Neo-Weberian contributions

The neo-Weberian housing literature has been dominated by Dunleavy and Saunders (Dunleavy, 1981; Saunders, 1990), the housing work of both having been heavily influenced by their wider interest in urban analysis, and particularly urban politics (Dunleavy, 1980; Saunders, 1979).

Dunleavy's main foray into housing was his detailed study of the decision-making process surrounding the introduction of high-rise council housing building and its subsequent demise (Dunleavy, 1981). The subject is treated as a case-study in the rise and fall of a political issue 'to analyse and understand the complex influences and processes which lay behind a particular series of policy-outcomes.' (Dunleavy, 1981: v). This is a detailed and exhaustive study, in which analysis of the state apparatus, both centrally (Part 1) and locally (Part 2), is a major – even dominating – concern. It includes both a preliminary examination of the politics and administration of public housing and three case-studies of local authorities and their handling of the high-rise issue.

Dunleavy, from a background in the discipline of political science, attempts to integrate his housing research closely into the wider concerns of the political science debates surrounding the question of power in society, in particular pluralism, 'new pluralism', élite theory, and Marxist theory. Both the national and the local government sections of his discussion are therefore concluded with conceptual chapters relating the findings to these debates.

National government, Dunleavy concludes, can best be understood in terms of the neo-Marxist perspective, 'stressing the political power of private capital' (Dunleavy, 1981: 191). He does, however, contend that new pluralism, or the post-industrial society thesis, and the arguments about the autonomy of technocratic decision-making, has 'impressive' descriptive accuracy, but 'remains flawed in its lack of reference to structural processes and influences' (p. 191).

Dunleavy therefore takes an essentially non-statist position concerning power and the national state in terms of housing policy. But he places considerable importance on the autonomy of the technostructure (both of the state and other institutions), and to that extent has elements of a potential statist perspective, even though this remains undeveloped and embryonic.

Turning to the local state, Dunleavy concludes in his final chapter 'that local authority decision-making on the high-rise issue was more determined than determinant, and that explanations in terms of structural pressures and influences substantially account for the development of local policies' (Dunleavy, 1981: 351).

Dunleavy's position may be summarised as follows. He sees the main influences on housing policy in the high-rise issue as being of a class nature, and by and large he takes a non-statist perspective on decision-making in policy. At the same time, he is sensitive to the arguments concerning the relative autonomy of the technocracy. What he does not do, however, is to place the analysis in the context of a clear statist thesis.

This is disappointing, since Dunleavy devotes considerable space to a careful analysis of the formal governmental decision-making machinery, both at the national and local levels. He devotes an entire chapter to a detailed description of the national political housing construction decision-making process that is unsurpassed in the literature (Dunleavy, 1981: ch. 1). This lends itself much more naturally to an evaluation in terms of statist theory than to the predominantly non-statist perspectives such as pluralism, élite theory and neo-Marxism, but Dunleavy never makes the direct connection necessary to accomplish this.

But one must understand the tendency to draw upon non-statist perspectives in terms of the very shape of the political science debates at the time when Dunleavy was carrying out his research. The 'Return to the State' thesis in political science really started only during the first half of the 1980s. Until then, the predominant perspectives were primarily 'non-statist' in orientation. Dunleavy's analysis was therefore principally reflecting the currently dominant perspectives in political science.

Saunders' analysis of the role of the state is primarily developed in the context of his work on urban sociology and the sociology of consumption, rather than in his direct contribution to the housing literature. In *Social Theory and the Urban Question*, Saunders develops what he terms 'a dualistic theory of politics' in which a different logic operates for the politics of production and of consumption (Saunders, 1986: 306).

This dualism has important implications for the way in which Saunders views the state. He argues that the modern state is highly complex with a whole range of different agencies and sub-units and that these different parts operate according to different logics and are influenced by different social groups, ranging from class interests to state employed officers and professionals. Because of this, no single general theory can be adequate enough to explain the way in which the state operates and which interests are represented through it (Saunders, 1986: 305–6).

This position suggests that a number of theories of the role of the state are required, including both statist and non-statist elements. However, Saunders does not elaborate on this. He argues for a dual politics thesis, roughly coinciding with the division between the politics of production and the politics of consumption. According to this, the politics of production are mainly carried out through the central state in which the mode of interest intervention is primarily corporatist and can be understood in terms of class theory. The politics of consumption, by contrast, can be better understood in terms of the local state, a plurality of consumer interests operating in a competitive mode, and through the application of interest group theory, or imperfect pluralism. Saunders' view of the state therefore divides into principally a Marxist approach at the central state level and a pluralist approach at the level of the local state.

It must be borne in mind that Saunders is not primarily interested in developing a theory of the state. Rather, he is concerned to develop a sociology of consumption, in which the role of the state must be understood, but only as one piece, albeit important, in the larger puzzle.

It is therefore not surprising that the explanation of the role of the state in urban politics that Saunders develops is only sketchy and in parts is unclear. In particular, precisely what Saunders means by 'corporatism' in the central state and 'imperfect pluralism' in the local state is open to question. More important, for our purposes, is that both elements of the dualistic theory of politics are based upon essentially a non-statist understanding of the exercise of power. Saunders does not develop a managerial thesis, or try to argue that the state can or does operate in society in any way independently from the social groupings – whether classes at the national level or interest groups at the local level – that are the ultimate sources of societal power. To that extent, therefore, Saunders views the role of the state as largely mediatory and his treatment of it remains conceptually unexplicated.

We may conclude this brief overview of the literature as follows. It is clear that in both the Marxist and non-Marxist literature there is a heavy bias towards a view of the state as being largely, if not entirely, the passive tool of wider societal interests, with, at best, mediating functions. Researchers concentrate on analysing the ways in which societal groups can bring their interests to bear on policy by using the state as a means to this end. This is true of all of the works discussed above, though some are more explicit about the relative insignificance of the state than others. Nor is there much attempt by housing researchers, with the important exceptions of Dunleavy and Saunders, to draw upon theories of power in the political science literature in any systematic way in order to provide the basis for their analyses. At the same time, there is considerable awareness of the potentially independent role of the state, particularly in the work of Lambert *et al.* (1978) and Dunleavy (1981). This, however, is always seen as subsidiary to class theory explanations, and is never analysed explicitly in terms of any theory of autonomous state power. Nor is there any attempt made to weigh up the relative importance of autonomous state power on the one hand and societal interests on the other.

THEORISING THE STATE IN HOUSING RESEARCH

An important element in developing a better conceptualised housing studies must be to begin to explicate the kinds of position that the state may be seen to have in relation to housing issues. I do not propose here to propound a 'statist' thesis, except in so far as I argue for a less societally reductionist approach to the state than has been current in housing research. The 'political reductionist' versus 'societal reductionist' debate in political science represents a polarisation which is useful as a heuristic device in refocusing researchers' attention on the relatively neglected issues of autonomous state power (Almond, 1988: 872), but it is fairly clear that there can be no return to the state in any simplistic political reductionism which ignores or neglects wider social forces and their operation through the state to determine or influence housing policy.

Rather, a new balance must be found, based on an awareness of the issues raised in the political science debate over the return to the state. Such a balance must retain the advantages of existing perspectives in housing studies which are sensitive to non-state sources of social power, but must introduce a clear understanding of the extent of the power of the state, as an autonomous force in relation to housing issues, which is not reducible to classes, pressure groups, vested interests, or whatever.

A central task in housing research must therefore be to begin to develop an understanding of the role of the state in the housing market in relation to vested interests both in housing and in wider society. I want to argue here that it is inadequate to assume that state involvement in housing is generated as merely the institutional arena within which the interests of power groupings in society bring their interests to bear. Instead, it is fruitful to understand housing policy in terms of interaction between wider social vested interests on the one hand and the practices and interests of the state on the other.

A good example is the question of the political support for policies that encourage owner occupation. Most observers imply that this is a reflection of societal interests which force governments to support and encourage owner occupation. Ball *et al.* (1988: 17–18), for example, argue that the growth of home ownership after 1945 was due primarily to rising real incomes, but ultimately 'all the major policy initiatives with respect to home ownership have to a great extent been forced on governments because of the developments in its structure of provision' (Ball *et al.*, 1988: 31). Saunders (1990), by contrast, suggests that state support for home ownership is largely a reflection of popular demand.

On the face of it, both of these views seem incompatible. However, they have in common a view of the state as essentially reactive. For Ball the state reacts to vested housing interests. For Saunders the state reacts to the preferences of electors. But such passive views of state intervention are unsatisfactory. In the first place, it remains unclear just why the building industry, housing finance institutions, etc. should prefer owner occupation to other forms of tenure, since new houses cost the same to build and finance irrespective of which form of tenure they are sold in. The ontological security argument is also inadequate since it conflates ontological security with a culturally specific form of tenure (for a critique see Gurney, 1990). Second, a case can be made out for a largely political explanation of the state sponsorship of owner occupation, on at least two levels. The first is that of legitimation: that governments see considerable electoral advantage in loading the housing dice in favour of one form of housing as against others. The second is a broader ideological one in terms of the beliefs and assumptions of politicians and senior civil servants and the generation of 'myths' which provide basic emotive power behind which support can be mobilised for public policies (see Chapter 7).

In this view, the state has its own reasons for supporting owner occupation and is a major catalyst in entrenching and reinforcing its popularity. Council house sales in Britain is a good example of this, which would be hard to explain in terms of vested interests and/or demand by council tenants for sales, without a major explanatory element being devoted to the political process.

A strong case can therefore be made out for analysis of public policy in terms of the interaction between the interests of powerful social groups/ classes/organisations or whatever, on the one hand, and the interests and power of those responsible for the political and legal institutions of the state on the other. The latter may be diverse and complex. For example, the electoral interests of governments may be supported – or undermined – by the beliefs, interests and organisational goals of the administrative arm of government. There are also major cleavages between central and local government, as many housing researchers are aware, and while much can be learned from detailed case studies, such as Forrest and Murie's (1985) study of the impact of the 1980 Housing Act on local governments, these need to form the basis of theorised accounts of the relationship between central and local states.

All of this suggests that policy analysis requires a more sophisticated approach in which private and public interests interplay to produce policy outcomes and different patterns of state intervention (for a useful recent attempt at this see Rothstein, 1990). Furthermore, public interests may be as diverse and complex as private ones. The state cannot be seen as a simple monolith, as Saunders rightly argues, but as a complex amalgam of a wide range of interests and pressures, some of which may be conflicting and contradictory, both at the central and local level.

CONCLUSIONS

Returning to the state in the context of housing research must involve first and foremost a much more adequate contextualising of housing issues in the broader debates in political science, including Marxist theory, than is currently the case. It should no longer be good enough for any study of housing policy to be undertaken without first explicating a position on the extent to which the state is seen as an autonomous actor in the housing market.

Second, housing researchers need to begin to debate actively the relative weight of non-statist and statist perspectives in explanations of housing questions. Only through such an engagement can housing researchers begin to develop particular theories of the role of the state in housing and begin to see emerging an interaction between the findings of empirical housing research and the refinement or reformulation of theories and concepts of social power and the state.

Third, housing researchers need to become more sensitised to incorporating theories of the state into their empirical work. This means going beyond prefacing an empirical study with a review of the social sciences literature on the state, or even a conceptual chapter. It means designing research from the beginning which integrates theoretical questions with empirical and policy issues. Because of the centrality of the state in relation to housing issues this will necessarily go a long way towards developing more general conceptual awareness in housing studies.

4 A critique of unilinealism in comparative housing research

INTRODUCTION

Explanations of social change lie at the heart of social science research. The very origins of sociology are to be found in the profound social upheavals that took place in Europe at the end of the eighteenth century in the form of the French revolution and the early stages of British industrialisation. Before this time, the relatively slow rate of change of social structure was itself a hindrance to the development of sociological consciousness. With the enormous changes that have engulfed society, every major social analyst from Comte through Marx and Weber to the present have grappled with understanding the structural factors at play in forming modern – and 'postmodern' – society. In this sense, social change constitutes the central focus of social theory.

Comparative housing research has hitherto tended unquestioningly to accept the dominant theoretical position in most of the social sciences that addresses the issue of social change in terms of theories based on unilinearity and the convergence of social structures. Underlying comparative housing research can be found an unexplicated and often unconscious use of unilinear theories of social change based on concepts of 'development' and 'evolution' in the Social Darwinistic tradition. In this chapter, I criticise this unconscious adoption of evolutionism by housing researchers as a starting point for subsequently theorising comparative housing in terms of multilinear change and divergence, developed further in Part III.

The main issue I address can be stated as follows. While there exist theories which explain similarities between societies, often in the form of basic underlying processes of social change common to industrial – or sometimes more narrowly capitalist – societies, there is a lamentable lack of theorising around the question of how to explain divergences and varying trajectories of social change. Theories of social change typically focus upon shared elements in societies and the tendency of societies to conform to a common trajectory of change even when societies differ from one another quite markedly. Such theories therefore produce explanations of social change which build upon observed uniformities. In so doing, they concentrate on highlighting similarities between societies and disattend to differences. There is therefore a tendency for explanations of social change to be made in terms of unilinearity and convergence rather than multilinearity and divergence. Societies are seen as moving in one direction that can be predicted from some set of theoretical postulates. Differences between societies tend to be explained, if at all, in terms of different stages of development along the same basic trajectory (for example, in terms of 'underdevelopment' or 'maturity'), or in terms of variations upon a basic and common structure, or else in terms of specific and idiosyncratic – and hence essentially non-theorisable – historical circumstances.

My concern with divergence rather than unilineal change reflects a basic departure from dominant evolutionary 'developmentalism' of both Marxist and non-Marxist varieties. It represents an abandonment of the search for an integrating common factor in industrial society and focuses instead upon understanding the extent to which differences between societies can be theorised. It should be noted here that I do not claim that developmental theories are wrong in some absolute sense. It is just as legitimate to neglect major differences between societies as it is to neglect underlying similarities. In one sense, the difference may be conceived in terms of different levels of analysis. Focusing on the process of industrialisation at a general level does tend to minimise differences between societies. My interest, by contrast, lies in 'middle range' theorising at a level that hitherto tends to be explained away in terms of residual factors.

Here I therefore begin the task of theorising the process of divergence in industrialised societies with special reference to housing by first reviewing the existing literature and developing a critique of its basic unilinearity and convergence orientation. The discussion is divided into two parts. First, and by way of introduction, I discuss the background to unilinearity and the convergence thesis in the context of general theories of social change in the social science literature of the early postwar period. The main part of the chapter comprises a critical review of the comparative housing literature with the aim of showing that unilinear theories of social change and convergence, which explain away differences between societies as mere variations or as uneven development, have had a pervasive – almost hegemonic – influence on explanations of differences between national housing systems.

This chapter must be seen as a ground clearing-operation before the attempt in Chapter 6 to develop the beginnings of a 'divergence thesis' in terms of housing. I argue that we must begin to move away from unilinear and convergent explanations of change, which treat differences between societies as being theoretically residual, and begin to think of differences between societies as constituting a central problematic in its own right, with the aim of developing theorised explanations of them.

UNILINEARITY AND CONVERGENCE IN SOCIAL THEORY

The convergence thesis has had a profound influence on social science research, and particularly on comparative research. The basic ideology of social evolutionism, which dominated both Marxist and non-Marxist thinking during the nineteenth century and to a large extent continues to do so today, reflects the underlying belief in the fundamentally evolutionary nature of social change – in 'development' and 'progress' (Wallerstein, 1984). That belief was not limited to thinking in the social sciences, it derived from much older intellectual roots in religious philosophy that placed the human race at the pinnacle of life, a belief given scientific legitimacy by the work of Darwin.

The ideas of unilinearity and convergence have therefore had a powerful, if often unexplicated – and therefore concealed – influence upon comparative research. Both Marxist theories of the maturation of capitalism, and non-Marxist theories of post-industrialism and of the end of ideology, build implicitly – and sometimes explicitly – upon the idea that industrialised societies share a common trajectory of social change. Societies may be at different stages of development and may exhibit unique characteristics which reflect specific – and therefore untheorisable – historical circumstances, but, ultimately, explaining social change centres on developing theories which account for the shared experiences of advanced industrialised societies.

The convergence thesis is associated most clearly with Kerr *et al.* (1960). Their use of the expression 'the logic of industrialism' has been widely cited as the dynamic leading to the gradual homogenisation of industrial societies. Among those who have interpreted Kerr *et al.* (1960) in this way are Goldthorpe (1964) and Giddens (1982). Yet a closer reading of the now classic text of Kerr *et al.* shows that such a view is based on a misreading. They were by no means arguing that industrial societies *inevitably and necessarily* become increasingly similar to one another. Rather, they argue that social change takes place in interaction between on the one hand what they call 'the threads of diversity' – deriving from a combination of cultural variations and the strategies of élite groups – and on the other hand 'the sources of conformity' comprising the logic of industrialism. They argue not that the logic of industrialism will overcome the sources of diversity in cultural and social structural variation but that the process of social change involves a constant struggle between these, the outcome of which is essentially unpredictable. This is made abundantly clear when they conclude their book with the following:

> Pluralistic industrialism will never reach a final equilibrium. The contest between the forces for uniformity and for diversity will give it life and movement and change. This is a contest which will never reach an ultimate solution... The themes of uniformity and diversity, and manager and managed which mark the world today will characterize it in the future as well. There will be constant adjustments between these eternally conflicting themes, but no permenant settlement.
>
> (Kerr *et al.*, 1960: 296)

It is quite clear that Kerr *et al.* (1960) were *not* arguing that industrial societies are necessarily and inevitably converging as a result of the logic of industrialism. However, their whole approach and the context of the sociological debates within which they must be understood was one of countering the powerful Marxist theory of the mounting crisis of capitalism. What they shared with Marxism was a belief in an underlying 'logic' that pushes forward social change. A similar concern is reflected in the work of Rostow (1960). His attempt to describe stages of economic growth was an explicit counter to Marxism, as its sub-title clearly suggests. He placed societies along a continuum of industrialisation culminating in what he termed 'the age of high mass consumption' and attempted to quantify the temporal gap that separated countries along this trajectory. But both Rostow (1960) and Kerr *et al.* (1960), as well as their Marxist opponents, were still basically concerned to describe what they understood as the common trajectory of social change of industrial societies.

It is therefore not surprising that sociologists have identified such work as being key contributions to a convergence thesis, especially since they themselves were equally engrossed in the same evolutionist task. The belief that Kerr *et al.* (1960) is therefore one of the key texts positing the convergence thesis has become firmly entrenched in social science folklore, while the nuances of the argument have been forgotten or neglected.[1]

The fact that the convergence thesis has proved to hold a fascination for researchers basically concerned with the same problem of unilinearity is strengthened by the search for theoretical simplicity in explanations of social change. If it is possible to find one basic dynamic that underlies social change, then the process of industrialisation can be explained in terms of that dynamic or 'logic'. This in turn means that industrialisation is predicated upon a powerful tendency towards uniformity in social structure. Built into most theories of social change, then, lies an implicit unilinearity and very often also an implicit convergence thesis.

Theories which attempt to understand processes of change in terms of what might be termed 'divergence' are much less common. Divergence implies a multiple branching trajectory of social change in which different dynamics operate to produce different social structures. Such a theory is clearly more complicated to develop than simple unilineality, since it requires an 'if not A, then B' logic.

It is therefore not surprising that theorising diversity – even simple 'one-of-two-alternatives' divergence – is rarely, if ever, attempted. Explanations of differences are normally interpreted in terms of 'variations' on a basic theme, variations produced by specific historical circumstances. A prominent example of this is Barrington Moore (1966) in his description of how different forms of industrialisation resulted in different outcomes in terms of political democracy, depending on a variety of very specific historical factors derived from the manner in which lord–peasant relationships happened to have evolved.

This 'variations' approach is well exemplified by Marxist theory, based on the dynamic of class conflict and the process of capital accumulation and the falling rate of profit. This is clearly a unilinear theory. It is also a convergent theory in so far as the end state of communism is common to all societies. For this reason, Marxist explanations of different patterns of development are commonly advanced in terms of variations owing to specific historical circumstances and often also some concept of 'uneven development'.

THE CONVERGENCE THESIS IN COMPARATIVE HOUSING RESEARCH

As with so much else in housing research, the issues of unilinearity and convergence, which arouse such considerable theoretical interest in the social sciences, have been all but completely ignored by housing researchers. Comparative research on housing has often been predicated on implicit theories of unilinearity and convergence, and has sometimes even been based on what might be termed a 'covert' convergence thesis.

The comparative housing literature is very substantial, and it is not the intention here to provide a comprehensive overview of the treatment of comparative housing research in terms of unilinearity and the convergence theses. Rather, the aim is to select a number of key works supplemented by my own work in this area and indicate how these have handled unilinearity and convergence.

The books considered here are Donnison (1967) – together with its updated version (Donnison and Ungerson, 1982) – Castells (1977), Ball *et al.* (1988) and Dickens *et al.* (1985). The first is chosen primarily because it and its update are prominent and influential examples of non-Marxist work in housing. The second is chosen because of its prominence and influence as an early Marxist housing analysis. The third and fourth are chosen as recent Marxist contributions which both explicitly set out to theorise comparative research on housing.

Donnison and the drift to comprehensive policies

Schmidt (1989a) tests Donnison (1967) and Donnison and Ungerson (1982) for the convergence thesis he argues that they embrace. According to Schmidt, the influence that explicit convergence theory has had over housing research:

> can be traced to one study, namely David Donnison's *The Government of Housing* published in 1967. In this book Donnison develops the thesis that the housing policies and the housing markets of industrial societies are converging – irrespective of party-political, ideogical, or institutional circumstances – because the societies are developing towards increasing economic and demographic similarity. In a follow-up of this study, Donnison modifies his position (Donnison and Ungerson, 1982). Nevertheless, the basic thesis remains unchanged: economic development accompanied by specific demographic changes leads to the convergence of housing policies and housing markets of industrial societies.
>
> (Schmidt, 1989a: 2–3)

A closer reading of the two books concerned shows that Donnison does not posit an explicit convergence thesis. Nowhere does he draw upon the social science literature on convergence or refer to the debates around this issue that took place within sociology and political science. He does argue that housing policies tend to follow a 'developmental' pattern towards more 'comprehensive' government involvement in the housing market (Donnison, 1967: ch. 3; Donnison and Ungerson, 1982: ch. 5). However, this is heavily qualified, especially in the later book, as a tendency, particularly in the light of the trend towards governmental disengagement from involvement in direct housing provision and direct subsidies – as distinct from tax deductions and exemptions (Donnison and Ungerson, 1982: ch. 1).

Schmidt therefore exaggerates the theoretical explicitness of Donnison's work. At the same time, it is clear that Donnison writing in the mid-1960s would have been influenced by the debates being conducted in the social sciences over the convergence thesis by Kerr, Bell, and others (see Goldthorpe, 1964; Dunning and Hopper, 1966), just as in the late 1970s he and Ungerson were to be influenced by the trend towards ideologies of governmental disengagement from housing that were becoming increasingly strident at that time.

The position that Donnison adopts is more accurately described as an unexplicated and covert convergence thesis of the sort that Schmidt ascribes to other housing researchers. Donnison can therefore be placed firmly in the traditions of atheoretical housing research, in that he does not refer to the existing theoretical debate, nor does he theorise in any explicit sense about either unilinearity or convergence in terms of housing policy and the housing market. Rather, he links the three types of housing policy models that he posits – haphazard intervention, 'social' (or 'residual') policies, and comprehensive policies – to different stages of economic growth; the first to the least 'developed' societies of Spain, Portugal, Greece and Turkey (Donnison and Ungerson, 1982: 67), the second to countries such as the UK and Switzerland (pp. 74–77), and the third to countries such as Sweden and West Germany (p. 83).

The tradition of evolutionary thought, with the ultimate convergence of social structures as the end result, is therefore clear. The implication – and it is only that – is that countries pass through stages of economic and social development which largely explain why the pattern of housing policy differs between them. Thus, for example, talking about social housing policies, Norway and Sweden are characterised as having 'once had the same sort of system' (p. 75), while Ireland and Belgium 'still exhibit some of the same characteristics' (p. 75). The whole tone of the presentation is to suggest that there are certain general types of housing policies associated with certain stages of economic development, but without explicating a theoretical position on the matter.

This kind of presentation of an argument may be described as 'theorising by innuendo'. The convergence thesis is nowhere explicitly stated, yet can clearly be read between the lines. However, it is important to recognise (as does Schmidt, 1989a: 3) that Donnison's contribution is significant largely because, despite the above criticism, he succeeds in going further than most comparative housing researchers in placing his observations within some sort of a conceptual framework.

Castells and unilinear Marxism

Castells (1977: 145–169) deals with housing only as part of a wider treatment of the urban question (though this amounts to the equivalent of a lengthy essay). His main argument in that essay revolves around the inability of the private market to provide adequate housing and the relationship between this and profitability within the context of the reproduction of labour. Duncan (1981) has criticised the disjuncture between abstract theory and empiricism displayed by Castells, and particularly in Castells (1977). That argument need not concern us directly. Nevertheless, it is important to understand that Castells' theoretical framework, with its implications of crisis-management and evolving capitalist relations of production, forms the basis for a fundamentally developmental analysis.

The key arguments are presented on pages 156–8. There Castells argues as follows:

1 His data, drawn from the experience of France, are based on the experience of a country with an endemic housing crisis that cannot be solved by private interests, and one that is 'the rule in most capitalist countries' (p. 156)
2 Countries where this crisis is less acute, such as Britain, Sweden, West Germany, and Canada, are countries where earlier heavy state involvement had already eased the situation.
3 The process in France therefore 'has a general validity' (p. 156), differences only being marked by varying degrees of the success of state intervention, the analysis of which 'requires a socio-political study of each country' (p. 158).
4 The only country where private interests have succeeded in providing most of the residential building, and also to a markedly better standard than in Europe, is the USA. Five reasons are given as to why this should be so, including such factors as the high material standard of living and the existence of 'lebensraum' reducing pressure on land.

The thrust of Castells' argument is summed up in the following quote:

> The housing question in France is not an exception, but a typical case, within the developed economy, *at a certain phase in its evolution.*
>
> (Castells, 1977: 158; emphasis added)

There are a number of points to note about this argument. In the first place, the unilinear development outlined by Castells is in line with Marxist theory and the idea of 'phases of evolution' with its obvious Social Darwinistic overtones, of which it is a particularly clear example. Castells is not arguing explicitly that societies are in a process of convergence and that in time all societies will become more similar to one another. But he is arguing as follows:

(a) There is a general tendency for the state to have to intervene in the provision of housing because of the tendency for private capital to be attracted to more profitable sectors of the economy (as a result of the inexorable decline in the rate of profit as predicted by Marxist economic theory).

(b) In some societies state intervention is more successful than others.

(c) In one society, the USA, there are special historical factors that explain why the private market has successfully been able to provide housing for the masses.

However, the convergence-like nature of this explanation is clear. Departures from the norm are to be explained not by reference to a theory but, on the one hand, by carrying out specific 'socio-political' case studies in the question of understanding why state intervention in the housing market has been more successful in some societies than others and, on the other hand, by drawing up an inventory of special circumstances that explain why the USA is an exception to the rule of the need for state intervention. There is therefore a linear process at work which has much in common with the arguments put forward by Donnison, in terms of increased state involvement in the housing market in order to resolve the crisis that cannot be managed by private interests; once this intervention has attained a certain measure of success, however, there may then be scope for private provision on a larger scale.

Recent Marxist work on 'variability' and 'uneven development'

Castells' treatment of housing is, of course, a relatively early contribution to the burgeoning Marxist housing literature, and it might be thought that more recent developments in Marxist scholarship have produced explanations that go considerably beyond such a simple and unexciting analysis. Unfortunately, this is not so.

Comparative work in housing inspired by Marxist theory is relatively uncommon. Two recent examples are Ball *et al.* (1988) and Dickens *et al.* (1985). These give a good indication of the kinds of changes in Marxist housing analysis that have taken place since the early 1980s. Such changes can be summarised as a growing awareness of 'variability' between different capitalist societies, and an attempt to explain this in terms of 'uneven development'.

Ball *et al.* (1988) differ from Castells (1977) primarily in that the emphasis is not so much upon delineating the general way in which housing markets of different countries can be seen to be following similar lines of development in response to broader processes of capitalist development but upon describing in outline the extent of variations between countries in response to commonly experienced changing circumstances. They therefore interweave the themes of general trends and specific variability in housing provision. As they conclude:

However, while the cross-national focus of our work has enabled us to comprehend the *generalised* nature of the restructuring which is now occurring, it has also highlighted the considerable degree of *variation* in the manner in which each nation is experiencing these changes. As we have insisted throughout this book, structures and relations of housing provision are nationally specific.

(Ball *et al.*, 1988: 199; emphases in original)

The authors then go on to criticise some of the Marxist work 'which, oblivious to much historical and contemporary evidence, attempts to simplify and reduce the variation and complexity in structures of provision via the use of ideal constructs and universally applicable hypotheses' (Ball *et al.*, 1988: 200).

However, this awareness of variations in forms of capitalism is taken no further than a descriptive analysis. In particular, theorisation remains largely implicit and underdeveloped. Therefore, while neither unilinearity nor convergence is posited, underlying the analysis is a basic assumption that capitalist societies are passing through similar crises and problems which are being handled in different ways in different countries. This does not really take Marxist housing analysis much beyond Castells (1977) other than to shift attention from the shared experiences to the specific variations between countries. However, this very eclectic approach, combined with the undertheorised nature of the discussion, means that the perspective shares much of the pragmatism and policy-fixatedness of which the authors are critical in the 'liberal–interventionist' perspectives (Ball *et al.*, 1988: 10–16), a point noted by Schmidt (1989b) and by Oxley (1989).

The treatment by Dickens *et al.* (1985: ch. 1–4) of the differences between Sweden and Britain in terms of housing construction and land policy is a much more theoretically explicated comparative analysis, made more focused by concentrating upon only two countries rather than attempting to cover many countries. The authors choose to describe differences between these two societies in terms of spatial variations in capitalism and this partly reflects their backgrounds in the geography and architecture disciplines. Taking a Marxist perspective on capitalist society, the authors, like Ball *et al.* (1988), are critical of the attempt to 'read empirical events directly out of abstract theory' (p. 2). They pose the question, 'How is it that the political economy of capitalism produces important variations in housing provision?' (p. 2).

Their solution to this is to argue that variations in the provision of housing derive from the fact that structures of capitalism are not fixed and immutable but are subject to variation, deriving from specific historical and spatial circumstances, as social actors struggle within the framework of relations between capital and labour. These idiosyncratic elements the authors term 'the contingent links made by housing provision, in such societies, between the economy, civil society and the state' (p. 17).

Dickens *et al.* therefore attempt to explain variations between capitalist societies in terms of 'contingencies'. This is basically a rather similar theoretical position to that of Castells who, as we have seen, also argues that there are variations between societies deriving from specific historical circumstances. What marks out Dickens *et al.* (1985) from Castells (1977) and also in a different sense from Ball *et al.* (1988) is primarily the fact that the former carry out an explicit piece of comparative research designed to describe systematically differences between two capitalist societies, and that they discuss in some depth the theoretical problems associated with the way in which abstract Marxist theory leads one into simplified generalisations.

The empirical analysis of differences between Swedish and British housing provision that is presented by the authors is detailed and of great interest. They show how, for example, the construction industries of the two countries are based on very different strategies for profit-making. The authors argue that, in Britain, profit is mainly made from a combination of land development profits (including land speculation) and extra profits (subsidised by the state) through house prices while, in Sweden, profits are more directly based on productivity in the construction process itself. The results of this analysis are clear. A considerable degree of variation exists in the manner of housing provision between the two countries, and the authors succeed in specifying the nature of the variation in a convincing and detailed way that goes beyond a simplistic inventory of differences.

Unfortunately, the authors do not discuss the place of their work in relation to the convergence–divergence problematic. Instead, they see differences between capitalist societies as being explainable in terms of geographical and historical contingencies and therefore as not being amenable to explanation in terms of theory that is designed to account for differences *per se*. Marxist theory explains the basic form, while historical contingencies produce variations between states and between localities. The contribution of comparative housing research according to this view is therefore to provide detailed case studies whereby variations on a basic theme can be spatially delineated and described.

But there is one further conclusion that must be drwan concerning the comparisons made between Britain and Sweden. Over and above the authors' emphasis upon structural variability and despite the formal claim that this is the basic explanation for differences between capitalist societies, there remains an implicit Marxist unilineal thesis, concerning the differing trajectories of the development of capitalism in the two countries. This unilineal thesis is not argued for in the theoretical discussion but is derived by the authors from the conclusions they draw to explain why housing provision in Britain and Sweden differ so markedly, and what policy measures can be proposed to deal with this.

The authors argue that Sweden represents 'a purer form of capitalism' (Dickens *et al.*, 1985: 36), as a result of 'a relatively early, rapid, thorough and uncomplicated establishment of the capitalist mode of production' (p. 38), which is contrasted to Britain with its lingering and still strong aristocracy, finance capital interests and traditional working class aristocracy. It is in this 'purer form of capitalism' argument that the real explanation lies for the differences between the two countries. The authors contend that in Britain relations between capital and labour still retain feudal and early industrial elements that make land speculation the main source of house-building profit, in contrast to Sweden where profits are made from the construction process itself.

This evolutionary perspective is taken further in a later book by two of the authors, in which they posit a thesis of the 'uneven development' of capitalism (Duncan and Goodwin, 1988: 61–73), a thesis which fits well with the earlier comparative treatment of Britain and Sweden. In the final analysis, then, the treatment by Dickens *et al.* (1985) of the Swedish–British housing comparison rests upon a traditional Marxist 'stages of development' argument along evolutionist lines. The lessons for Britain are therefore primarily based on modernising its building industry along more rational capitalist lines, steered by strong state intervention. Such a prescription has clear convergence overtones in that the solution is seen in terms of accelerating British capitalist development.

CONCLUSIONS

The conclusion that must be drawn from the above overview of prominent contributions to the comparative housing literature must be that it suffers from a number of major defects. First, it remains heavily influenced – usually implicitly but sometimes covertly – by the evolutionism that has dominated social science thinking from its earliest days. This is true of both Marxist and non-Marxist approaches. Second, a latent and implicit convergence thesis can be read into at least some of the literature. Finally, some of the more sophisticated recent approaches to comparative housing research have demonstrated an awareness of the tension that exists between developmental, linear-like explanations of social change on the one hand and quite major diversities and differences between industrialised societies on the other. This tension tends to be resolved by explaining such diversity in terms of 'variations' on the dominant mode of industrialism (or capitalism, depending on whether a non-Marxist or Marxist perspective is being used).

What is required in order to move beyond unilinear and convergence-like explanations of social change in the housing literature is to begin to develop theories that are specifically designed to explain the differences between societies in terms other than 'variations' or 'stages of development'.

This brings me back to my own work on comparative housing; a work which suffers from all of the defects of 'theorising by innuendo' that have been criticised above, but which was intended to provide the basis for a theoretical reconstruction in terms of divergence and which contains within it the elements needed to begin such a task. The first step in such an enterprise is to reconstruct critically the argument in *The Myth of Home Ownership* (Kemeny, 1981) in order to explicate the divergence thesis which underlies it. But before going on to do this it is necessary to complement the overview of theories comprising this section by examining the contribution of comparative welfare state analysis to theories of social change and by considering the relationship between welfare and housing.

5 Housing and comparative welfare research

INTRODUCTION

Housing has a peculiar position in the welfare state. The principle that underlies the welfare state is generally that of universality in a number of key areas. The major areas of welfare – education, health, unemployment benefits, retirement pensions – are all based on the principle that they are automatically available to everyone who fulfils certain minimum conditions of age or incapacity without means or needs testing. Housing is different in that the principle varies with different forms of housing, and in particular in terms of tenure. Owner occupation is, contrary to popular belief, subsidised on 'welfare' principles in that access to subsidies is not means or needs tested. Receipt of state subsidies is automatically triggered by the act of borrowing to buy a house. Public housing, by contrast, is not provided on the same universalistic welfare principles, assumed to be available to all, but is only available for a minority of the population, and only then after stringent, not to say draconian, screening and needs testing[1]. This special status of public housing has laid it open to potential political attack in a way that is not the case for public education or health or indeed owner occupation, to which access is not means or needs tested. As a result, housing – and in particular public rental housing – is, to use Torgersen's (1987) expression, 'the wobbly pillar under the welfare state'.

But, in whatever form it comes, housing in industrialised societies is inextricably bound up with the welfare state. Irrespective of how well or poorly developed the welfare state is, this is true to the extent that housing is subsidised directly and indirectly in all societies. In so far as there are differences between societies these tend to be in terms of what kind of housing is subsidised and how it is subsidised. Furthermore, it would seem that, irrespective of the way housing support is organised, the lion's share of support for housing goes to the better off rather than the poor. These 'laws' need to be addressed as fundamental theoretical questions in their own right, and it is an indication of the extent of the continuing poverty of housing research that they remain almost entirely neglected.

But the issue is much wider than that of 'state subsidisation'. We have noted in Chapter 1 that social administration has undergone something of a transformation since the early 1970s, with a rapidly growing theoretical awareness greatly exceeding that within housing studies. Welfare is no longer conceived narrowly in terms of statutory provision but is defined as a service which can be provided by many institutions, including for-profit organisations and, crucially, the family.

This has yet to make any impact on housing research. The intimate relationship between housing and welfare, and the rapid theoretical developments in welfare research, point to an urgent need to begin to explore the relationship of housing to welfare in general and theoretical terms. This chapter is directed to that end and takes the form of a preliminary ground-clearing operation. It takes a comparative approach, drawing on the burgeoning international comparative literature on welfare. The discussion begins with an overview of the principle theories of change espoused by welfare research. It then moves on to discuss some of the recent work on comparative welfare which begins to move away from traditional conceptions of the welfare state. Finally, the relevance of this recent work to housing is discussed, with a view to sketching out a preliminary conceptual framework for relating housing to welfare.

THEORIES OF CHANGE IN WELFARE RESEARCH

Three models of social change – for the most part implicit and unexplicated – can be identified in comparative welfare research: developmental, cyclical/episodic, and divergent. The models represent different levels of scope of explanation. Developmental models tend to be concerned with large-scale historical dimensions of change, highlighting similarities in otherwise diverse societies. Cyclical/episodic models tend to be based on cycles of social, political, and above all economic, activities and present a non-development 'history-repeats-itself' view of social change. Divergent models emphasise that the shorter term, often political, swings in fortune that can be discerned can, under certain circumstances, be cumulative, leading to significant divergence between societies.

It is important to note that the different models are not necessarily mutually exclusive or contradictory. It is quite possible to argue that at different levels of explanation different questions are being asked. Developmental models focus on long-term fundamental structural changes, which do not necessarily deny that either shorter cyclical changes or political differences and emphases exist within welfare programmes. It is simply that the larger scope of the questions being asked necessarily involves disattending to the shorter term fluctuations and differences, or vice versa. The mere fact of a more detailed focus does not necessarily deny the validity that more profound and far-reaching structural changes underlie it (for a discussion of theories of social change see Boudon, 1986). The argument here, as in Chapter 4, is not that unilinear theories are necessarily wrong, only that there is also a major place within housing research for theorising differences as well as similarities between societies (for a useful summary of this literature see Wilensky *et al.*, 1985).

The other major theme to run through this comparative welfare research is the role played in determining its pattern by ideology. Theories of social change are not merely interested in describing change, they wish to explain it. And it is not a coincidence that ideological differences have constituted a central feature of such explanations, albeit with very different conclusions being drawn by different observers.

Developmental: Wilensky and economic development

Developmental models of welfare often assume a dominant engine of social change, such as 'the logic of industrialism' that was observed in Chapter 4. This model is the most common of the directional models, and is reflected in Wilensky's (1975, 1981) argument that all industrial societies, irrespective of political ideology, develop welfare states largely in response to economic growth and demographic change. O'Connor (1973) and Offe (1984) represent a Marxist variant of this kind of model, in which the welfare state is seen as developing in response to capitalist dynamics, but then reaches a 'fiscal crisis' stage. 'Modernisation' models based on the posited relationship between citizenship and welfare, which develop in harness with one another, are yet another variant of developmental models of social change (Bendix, 1964; Marshall, 1950; Rokkan, 1970).

One of the most influential empirical cross-national studies of the welfare state has been that by Wilensky (1975). Using data on a large number of First, Second, and Third world countries, he conducts regression analysis on a number of variables in order to develop inductively a model of explanation for the development of the welfare state. Wilensky finds that the most important variables explaining the development of the welfare state are per capita GNP, the percentage of the population that are over 65 years old, and the length of time that the social security system has been in operation (inertia). His analysis discounts ideology as a major independent variable.

Looking at a small group of the top 22 industrialised societies, Wilensky speculates upon the reasons for the 'impressive' variation in social security spending as a percentage of GNP at factor cost in 1966 (the last full year of his data), ranging from 21.0 per cent for Austria and 19.6 per cent for West Germany to 7.9 per cent for the USA and 6.2 per cent for Japan: the differences between 'leaders' and 'laggards' (Wilensky, 1975: ch. 3). He concludes that the main differences are largely political and social class in nature:

> In sum, among rich countries, the welfare state will be most developed and supporting welfare-state ideologies most powerful where a centralized government is able to mobilize and must respond to a large, strongly organized working class with only modest rates of social (occupational and educational) mobility; where the middle mass does not perceive its tax burden as grossly unfair relative to that of the rich and the upper middle class and does not feel great social distance from the poor; and finally, where the tax system has a low visibility, self-employment experience is meagre, and the private welfare state is limited.
>
> (Wilensky, 1975: 68)

Wilensky followed up these ideas with a test of the impact of political parties on the extent of welfare effort in 19 industrialised countries (Wilensky, 1981). He compares broadly defined 'left' and 'catholic' parties between 1919 and 1976 in terms of their political influence according to various indices, such as number of years in government, control of key positions, continuity of office, and numbers of years in office. This is a complex study, not least because Wilensky is testing the corporatist thesis of power in addition to his interest in the effect of political orientation on welfare. His conclusions are that left parties are less important than catholic in determining levels of welfare achievement. But his approach is flawed by the complication of corporatist theory, and more fundamentally by his extremely general definitions of left and catholic. In spite of this, Wilensky's work remains one of the most important comparative empirical tests of labour movement theory, even if he concludes on somewhat shaky grounds that political parties play a minor role in the development of the welfare state (Wilensky, 1981). He ultimately espouses a developmental explanation of the welfare state, revealingly indicated by his use of the terms 'leaders' and 'laggards' to depict differences in the development of welfare programmes.

Cyclical/episodic: Esping-Andersen and 'policy regimes'

Strictly cyclical models of welfare are relatively uncommon. Two examples of this kind of approach are Piven and Cloward (1972) and

Ruesche and Kirchheimer (1939). Piven and Cloward argue that welfare in the USA has a mainly regulatory function, expanding in times of recession in order to dampen social unrest and contracting in times of boom to force welfare recipients back on to the labour market. Their model, while not strictly comparative in scope, is presented in terms of conceptual generalisation, with at least the implication that it is applicable historically and comparatively beyond the US experience. A rather similar argument is that by Ruesche and Kirchheimer, who look not at welfare in the narrow sense of social security provided by the state but at the prison system as part of the broader system of social control. Their thesis is that penal practice varies with conditions of economic boom and recession. In times of labour shortages there is a tendency for penal practice to exploit criminal labour in order to make up labour shortfalls in the wider economy. By contrast, in times of economic recession, criminal labour is withdrawn from economic activity.

Ruesche and Kirchheimer's theory is truly comparative. They illustrate the exploitation of criminal labour as a supplement to paid labour in terms of such phenomena as classical and medieval slave galleys, British colonial transportation and prison factories. By contrast, they argue that in times of recession criminal labour is withdrawn by the simple expedient of the more extensive use of execution, solitary confinement, concentration camps, low or zero productive work (eg. stone breaking, treadmilling) and other torture, incarceration or exile measures.

The point about cyclical explanations is that social change in terms of welfare is not a matter of inexorable and steady progress towards more welfare. Rather, it must be seen in direct relation to the state of the economy. Marxist theories of the welfare state as a concession to working class pressure and/or as a regulatory mechanism of capitalism have much in common with such a view of social change. However, more common than cyclical theories is the less determinant variation of seeing social change in terms of episodes.

The episodic perspective on welfare is more concerned with short-term political swings in fortune, often reflecting changes of government between political parties of different ideological persuasions. And here the emphasis shifts more towards patterns of social change which at least have the possibility of coalescing into divergent social structures. Mann (1987) criticises Marshall's (1950) theory of citizenship and welfare for being both Anglocentric and evolutionary, and posits a range of means of institutionalising class conflict, including fascism and socialism, arguing that if the liberal model has become dominant in Europe then this is the outcome of geopolitical factors, principally war, an explanation that he locates in the autonomy of the state thesis (see Chapter 3).

But a more widely researched example of an episodic/divergence model of differences between welfare states is labour movement theory. According to this, the extent to which the welfare state will be developed will be a direct consequence of the power of labour movements to control government and push through policies of developing the welfare state. Much of this will take the form of 'swings and roundabouts', the gains of one election victory being wiped out by the losses following the next election defeat. The degree to which the welfare state will be developed will, in the long run, be a reflection of the extent to which socialist parties are able to win and hold on to political power. We will consider the divergent variants of labour movement below. Here, I want to illustrate briefly the episodic perspective with Esping-Andersen's (1987) concept of 'policy regimes'.

Esping-Andersen (1987) states the episodic case explicitly. He argues that 'a given organisation of state–economy relations is associated with a particular social policy logic', and that 'policy regimes' can be distinguished:

> Each nation exhibits its own unique regime characteristics and, as a society historically evolves, it will pass through distinct regime changes and institutional realignments. Therefore, the social policy of any country is distinctive.
>
> (Esping-Andersen, 1987: 7)

Here Esping-Andersen is arguing for something close to the 'variations' perspective that is prevalent in comparative housing research and that was highlighted in the preceding chapter. However, more recently, Esping-Andersen appears to have espoused a clearly divergent approach.

Divergent: Castles, Esping-Andersen and political divergence

The approach taken in Esping-Andersen (1987) is not strictly representative of either his preceding or his subsequent work. He more accurately represents the divergent perspective in comparative welfare research by way of his attempt to understand the Scandinavian experience of welfare in terms of social democratic political power. In this, he is following the main traditions of labour movement theory, much of which focuses on the Scandinavian – and in particular the Swedish – experience as the example *par excellence* of labour movement divergent explanations of welfare (Castles, 1978; Esping-Andersen, 1985; Korpi, 1983; Rein *et al.*, 1987; Stephens, 1979). In Esping-Andersen (1985) he argues that the political dominance of social democratic parties is under threat because of the declining economic importance of the working class in relation to white collar workers. A new political alliance therefore needs to be built in order to avoid 'the decomposition' of social democratic parties.

More broadly comparative and statistical than much labour movement theory is the work of Castles (1978). Like Wilensky (1975), Castles chooses to measure ideological commitment to welfare through a party-political index. He uses somewhat broader indices of welfare than those of Wilensky: public revenues as a percentage of GNP, education expenditure as a percentage of GNP, and infant mortality rates. He uses two samples of countries: a larger one of 25 countries and a narrower one of 15 of the richest countries.

Using the smaller sample, Castles shows that right party dominance is associated with weak welfare state development. Concentrating on Scandinavia, Castles argues that left dominance is a product of a combination of a strong left and a divided and splintered right.

In a later analysis, Castles (1982) conducts a more complex analysis of changes taking place among 18 democratic capitalist countries between the early 1960s and the late 1970s. He uses a range of dependent variables as being measures of welfare (such as various measures of public expenditure, revenue, and transfer payments, spending on health, education, and income maintenance programmes), regressed against a range of independent variables made up of a number of socioeconomic factors (GNP, imports/exports, percentage of population over 65) and a number of political factors (notably trades union membership rates, right and social democratic party seats in parliament/cabinet).

Castles concludes on the basis of this analysis that: 'On the whole, the evidence provided by changing patterns of public expenditure in the 1960s and early 1970s appears to offer little positive support to the idea of convergence trends developing directly from the operation of economic forces' (Castles, 1982: 67). However, in support of Wilensky's argument, he does find some evidence of the importance of demographic trends.

But the crucial conclusion that Castles draws from his analysis is the importance of political parties for changes in welfare expenditure; that: 'partisan control of government is a major determinant of policy outputs' (Castles, 1982: 88). Castles argues that this does not mean that changes in electoral opinion are readily translated into changed policies. Rather, party policies reflect more fundamental class and interest group alignments. He concludes:

> Thus, to accept that the partisan control of government is an influential aspect in policy formulation is not necessarily to argue for an historical and non-structural viewpoint. Politics clearly matter in the sense that party political structures institutionalize class and interest cleavages and

make them continuingly policy relevant. But the nature of such cleavages can and does change, and a much longer time perspective than is inherent in a study of little more than a decade would suggest that these changes must eventually be reflected in new institutional structures, new ideologies and new policies.

(Castles, 1982: 88)

Castles therefore adopts a divergence thesis on modern industrial societies in terms of political parties and class mobilisation, rather similar to that of Korpi (1978, 1983) and Esping-Andersen (1985), although Castles is somewhat more sympathetic to corporatist theories of power than Korpi.

Unlike Korpi, and Esping-Andersen (1985), both Wilensky and Castles base their analyses on the statistical testing of hypotheses. The approach necessarily colours their conclusions. Statistical correlations of factors with one another provide much useful information on the workings of specific factors in relation to large problems such as changes in welfare over decades or generations. But the approach is necessarily piecemeal in that complex societal scale phenomena are broken down into constituent parts and indices are constructed for particular social, economic and political phenomena. There is a limit to how many variables can be juggled with in this kind of work; ultimately, the determining of correlation coefficients tends to encourage the development of explanations which are based on the factors that are more easily measured.

Interestingly enough, one of the main critics in recent years of the covariance analysis of comparative public policy has been Castles, one of its most accomplished practitioners. He has argued that the incremental addition to knowledge achieved by carrying out yet more sophisticated covariance analyses is subject to dramatically declining marginal benefits, and that a more qualitative, historical approach based on fewer case studies is now more potentially profitable (Castles, 1989).

I shall return to other aspects of Castles' work, especially his later, more qualitative, work in the context of a discussion on hegemony. To round off this discussion of divergence perspectives we need to consider the most recent work of Esping-Andersen, which is in many ways one of the most exciting developments since the work of Wilensky.

Esping-Andersen (1990) has argued that a process of divergence may be taking place between the welfare states of industrialised societies. Using a substantial body of quite sophisticated and sensitive comparative data he clusters welfare state variations by 'regime types': 'liberal' (USA, Canada, Australia), 'corporatist' (Austria, Germany, France, Italy) and 'social democratic' (Scandinavia) (Esping-Andersen, 1990: 26–9). Of these three regime types, the liberal and social democratic types are familiar from labour movement theory. The corporatist regime type is familiar from corporatist theory, but less so in terms of welfare state models. Esping-Andersen argues that it provides a distinctive welfare model because it involves considerable state intervention but in such a way as to reinforce existing status differences, whether based on class, ethnic groupings, or religion. He argues that three factors are of prime importance in determining which policy regime will emerge: class mobilisation (especially of the working class), class–political coalition structures, and the historical legacy of regime institutionalization (Esping-Andersen, 1990: 29). He concludes by tentatively positing divergence:

> Sweden, Germany, and the United States may very well be heading towards three diverse 'post-industrial' welfare-capitalist models. The conflict scenarios that we have sketched here may solidify into lasting structural factors. But then, it is also possible that events will occur or changes will be introduced that fundamentally alter the course on which these nations now seem to be headed.
>
> (Esping-Andersen, 1990: 229)

Esping-Andersen (1990) is potentially of major importance in that it represents a serious attempt, using comparative data, to show that the variety of modern welfare states need not be dismissed as mere 'variation' but can be understood in terms of diverging societal forms.

TOWARD A THEORISED CONCEPTION OF 'WELFARE'

We have noted that social administration is gradually working towards theorising the focus of the subject, by escaping from the narrow statutory definition of welfare and developing a concept of welfare in terms of the reproduction of labour, 'care', or some other broad concept which includes voluntary, private, household, and other forms of welfare provision. Comparative welfare research has yet to come to grips with this major development, as, indeed, the above review reflects. It remains the case that comparative welfare research is based ultimately on a state provision perspective on social welfare. However, this is beginning to change, and in this section I briefly review two contrasting approaches to explaining welfare in comparative terms which succeed in escaping from a narrow focus on 'the welfare state'.

Castles and 'the wage earners' welfare state'

Castles' early work, as we have seen, has been directed to uncovering the reasons for a close relationship between social democratic politics and the welfare state. In Castles (1985) he takes a dramatically different approach. The question he poses is as follows. Why is it that, both Australia and New Zealand in spite of having strong labour movements whose Labor Parties consistently obtain a high share of the total vote, have not developed a strong welfare state and are, in Wilensky's developmental terminology, 'laggards'?

Castles is critical of much comparative policy research, including his own earlier work, for over-concentrating on direct political power, and neglecting broader issues of class politics (Castles, 1985: 3–4). He argues that, for specific historical and cultural reasons, the Australian and New Zealand labor movements adopted a very different political strategy to that of labour movements in Europe. This was based not on expanding the social wage but on ensuring that wage-levels were kept above a minimum, which, in combination with low marginal tax rates, ensured that workers' standards were maintained.

Castles terms this strategy 'the wage-earners' welfare state'. He distinguishes it clearly from other 'laggards', such as the USA, by the 'existence of a statutory wage regulation system in the former which continued to provide a national level of needs-fulfilment below which the vast majority of wage-earners could not fall' (Castles, 1985: 103). Castles concludes:

> It is because the Australasian labor movement's reform strategy did, and still to a large extent does, include such a conception of a national minimum level of needs-fulfilment that we are, despite all the previous reservations, entitled to identify in Australia and New Zealand a unique model that might be described as the *wage-earners' welfare state*. This model is to be distinguished from the residual conception precisely by a strategy of creating a national minimum and from the institutional conception by the fact that the criterion of inclusion was status as a wage-earner, rather than as a citizen.
>
> *(Castles, 1985: 103; emphasis in original)*

Castles qualifies his argument by pointing out that the great increase in non-wage poverty in modern society, in the form of single parents, the young, and the elderly, calls the Australasian labor movement strategy into considerable question. Nevertheless, the argument has important implications for the way in which welfare is conceived. Castles' argument succeeds in casting considerable doubt upon the very idea that focusing upon the statutory provision of the social wage is at all satisfactory in understanding the degree of welfare in a society.

Rose and 'the welfare mix'

A radical departure from the traditional comparative policy literature is the collection of papers in Rose and Shiratori (1986), showing the wide variations in the sources of welfare provision in a number of different countries. Rose in particular attempts to broaden the concept of welfare to cover the entire range of goods and services available to a household, and he argues that the state is only one of several different providers of welfare:

> From the perspective of the family, its total amount of welfare is more important than the particular sources of that welfare. Therefore we must think in terms of the *welfare mix*, that is, the contribution that each of three very different social institutions – the household, the market, and the state – makes to total welfare in society. The state today is the most prominent producer of welfare, but it is not the sole source. The market is also a major source of welfare, in so far as individuals and households purchase welfare services, and many workers receive welfare benefits as a condition of employment. Historically, the household has been the primary source of welfare, and a substantial amount continues to be provided within the family.
>
> (Rose, 1986a,: 14; emphasis in original)

Rose argues that the family remains a major source of welfare provision but that, unlike state or market provision, family provision does not involve monetised exchange. The increasing share of state provision of welfare during the process of industrialisation therefore involved the 'fiscalisation' of welfare in its transfer from the family to the state and its consequent monetisation (Rose, 1986a: 21). It might be added here that the neglect of household-provided welfare builds in to social welfare analysis a strong gender bias, since overwhelmingly it is females who provide this form of welfare.

Rose develops a threefold distinction of types of welfare provision: household, state, and market. In another paper, Rose (1986b) attempts to quantify roughly this threefold distinction for Britain across the whole range of welfare provision, including food (mainly market and household), care of children and the elderly (almost entirely household), health (mainly state and household), education (mainly state), and so on. In terms of housing, he divides the main tenures into market (private renting), 'market-cum-household' (owner occupation) and 'state-cum-household' (public renting), and attempts to quantify roughly the household contribution to maintenance (Rose, 1986b: 86–7). Johnson (1987) complements Rose's threefold distinction by adding the voluntary sector to the list of welfare providers, to give a fourfold distinction between state, market, voluntary agencies, and the informal sector (including family and neighbour provision), so producing a mix that he terms 'welfare pluralism'.

This new, more complex conception of welfare is beginning to have an impact on international comparative research on welfare. Maruo (1986) uses the same threefold distinction as Rose to argue that Japan's low state provision of welfare does not reflect less total societal welfare, only different ways in which it is provided, with greater household and corporate provision. The importance of employer-provided welfare in the form of retirement, health, and other benefits is underlined by Pinker (1986) in a comparative study of Japan and Britain, and by Esping-Andersen (1990) in his discussion of welfare systems in countries such as West Germany and the Netherlands which he characterises as corporate because welfare is stratified by occupational groupings, ethnic or religious cleavages.

Finally, work is starting to be done on the interaction between different forms of provision. Szelenyi and Manchin (1987) discuss the interaction between state and market forms of welfare provision in state socialist societies in the context of their increasing marketisation. They argue that when one form of provision – state or market – dominates a society it will create inequalities which can be ameliorated up to a point by encouraging the expansion of the other form, but that after a certain point has been reached the secondary form will begin to generate inequalities which in turn can be ameliorated by encouraging the initially dominant form. This 'mixed economy' (Szelenyi, 1989) model of welfare, with each form being balanced against the other, can be seen as the beginnings of attempts to understand the way in which different forms of welfare affect each other and result in a dynamic relationship that remains as yet almost entirely unexplored.

Summarising this literature, it is clear that comparative welfare state research has been moving away from evolutionary models in terms of 'leaders and laggards'. With greater emphasis laid upon divergence, it has been particularly influenced by labour movement theory which in turn has been largely inspired by the Scandinavian model of the welfare state. At the same time, a more complex view of welfare is in the process of emerging, which is forcing researchers to re-evaluate exactly what is meant by welfare. This has been in part inspired by the lack of 'fit' with either developmentalism or divergence in terms of simple labour movement theory, as shown by the work of both Castles and Esping-Andersen. But it has equally been the result of the conceptual maturation of welfare research, and an attempt to develop a more complex concept of welfare.

All this, however, is very recent and the current state of the art is still one of piecemeal efforts at what might be termed the 'deconstruction' of traditional welfare research approaches. The end result is a bewildering array of alternative means of conceiving of welfare, as yet neither systemised nor with any clear understanding of the ways in which the different forms of welfare provision interact with one another to create distinctive societal patterns of welfare provision. At the same time, these conceptual developments are only now starting to feed through to comparative welfare research which, as we have seen, has until very recently largely remained wedded to the traditional conception of the welfare state.

It is clear that comparative welfare research is beginning to move away from simplistic analysis of 'the welfare state', and that, in the wake of this reconceptualisation of welfare, major reorientations of comparative welfare research are likely to take place in the near future. Such reconceptualisation is likely to result in an extensive body of research which highlights the uniqueness of welfare patterns in individual societies – a 'varieties' approach to comparative welfare. But it is also likely to lead to more emphasis upon patterns of divergence which go beyond quantitative measures of the extent of the welfare state explained in terms of labour movement theory, and which begin to theorise the interaction between different forms of welfare in more complex ways. This in turn has important implications for the way in which comparative housing research can develop. The intimate relationship between housing and the welfare state means that developments in comparative welfare research must be taken into account. There is already growing awareness of this fact, even if very little research has yet been carried out along these lines. Thus, for example, Harrison (1986) has argued for a social division of welfare approach within urban studies, and Forrest and Murie (1988) have discussed some of the implications of a 'social division of subsidies' approach to housing. In the next section I relate the above discussions of theories of change and conceptions of welfare to comparative housing analysis.

HOUSING AND WELFARE IN COMPARATIVE PERSPECTIVE

The developments that have taken place within welfare research over the last two decades have slowly and painfully moved social administration from having a narrow and unexplicated focus on statutory provision to developing a theorised conception of welfare. As noted in Chapter 1, housing research is still at the very beginning of that journey. It is possible that some concept of 'residence' might form the kernel of a future conceptualised housing studies. However, it is not possible at a stroke to make any dramatic advance, and I make no pretence at even trying. My ambition is much more limited. It is to relate housing issues to some of the developments in welfare – and particularly comparative welfare – research.

Housing and the welfare mix

Rose's (1986b) conception of the welfare mix applied to housing in Britain is limited to describing the general mix of welfare forms within each of the three major tenures of public and private renting and owner occupation. Of course, even within each of these tenures there exist substantial differences which would make for wide variations in the welfare mix. These include differences between mortgaged and unmortgaged owner occupation, local government variations in public renting, differences between public renting and voluntary sector social housing, co-operatives, tied housing, furnished and unfurnished private renting, and different leaseholds. To this must be added local variations, which can be very substantial. In addition, there are considerable variations which depend on the particular circumstances of households even if their housing conditions are the same, with wide variations between state, market, informal and voluntary sources of welfare.

It is here that a broader conception of welfare involving a mix of different forms of provision can be useful, as, for example, argued by Forrest and Murie (1988: 4). By understanding welfare in terms of different constellations of state, household, employer, voluntary, and market welfare, we can begin to unpackage housing into household and dwelling welfare. The permutations of household and dwelling welfare undoubtedly define one of the major societal fracture lines between households in terms of inequalities and access to resources. Indeed, one of the stumbling blocks to proposals for large-scale housing policy reform has been precisely the difficulty in unravelling housing from social security provision. To take an obvious example, introducing market-sensitive rent-pooling in public housing is severely hampered, if not ruled out, by the manner in which social security benefits are calculated.

This complex inter-relationship at the interstices of dwelling and household constitutes one of the central conceptual problems which needs to be addressed in housing research, the clarification of which would also be a major contribution to social welfare research. But, in addition to breaking down the welfare mix into more detailed elements, it is possible to broaden the perspective from the internal structure of housing tenures and to consider the ways in which differently constituted forms of tenure combine to produce an overall welfare mix of the housing system as a whole.

Thus, for example, it may be that two societies have similar sized public rental sectors but that one has a surplus extracting and market-determined rent-setting system, dependence on tenant maintenance and repairs, and private construction and finance, while the other has non-profit rents, interest-free co-operative finance, and construction and maintenance using publicly-owned organisations. The overall nature of the welfare mix in the two societies may differ radically from one another even though on a superficial level they appear to have very similar proportions of 'the same' housing tenures.

However, housing has wider social implications which need to be considered. As argued earlier, housing is not merely about dwellings but also about residence, which includes locality factors, such as urban infra-structure, and public, private and other facilities which lie beyond the dwelling but which complement dwelling facilities, thereby influencing their composition. Two examples of this interpenetration are the impact of dwelling type on urban densities and thereby on the nature of the urban transport system (see Chapter 7), and high levels of private washing machine ownership resulting in lower levels of local self-service laundry facilities.

This suggests that housing can be analytically broken down into a threefold distinction of levels of increasing scope of social structure, namely household, dwelling, and locality. These in turn can be understood in relation to different forms of welfare provision. That is, the welfare mix can, and usually will, vary widely between the three levels. Figure 5.1 illustrates the basic concept.

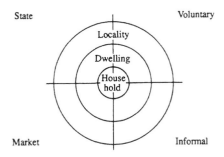

Figure 5.1 Dimensions of housing and forms of welfare

The three concentric rings of household (type, stage of family cycle, socioeconomic status, etc.), dwelling (type, size, conditions, facilities etc.), and locality (local complementary facilities, transport and communication, social characteristics of the neighbourhood, etc.) constitute *in toto* the phenomenon of residence and its impact upon social structure. Each dimension will be characterised by a particular mix of forms of welfare, with different degrees of importance of state, market, employer, voluntary and informal provision, varying between households, locally, nationally, and internationally.

Such a model could be a useful starting point for understanding divergent patterns of welfare. Societies with divergent patterns of welfare provision would manifest wide variations in the balance between different forms of welfare provision in all three dimensions of housing. We might also expect there to be some congruency between the balance of welfare provision in the dimensions, since they are overlapping and interpenetrating and closely affect each other. It should also be possible to say something about the dynamics of the mix and the way in which changes in one affect the others.

Housing and divergent welfare systems

The comparative welfare research discussed earlier in this chapter is remarkably silent about housing as an element of welfare. Wilensky (1975: 7–9) explicitly excludes housing from his analysis on the grounds that, while housing is an important dimension of the welfare state, there are formidable measurement problems. These, he argues, derive from a combination of the poor housing data available and the complex interplay of housing with other factors (such as monetary policies, rates of interest, roadbuilding and public transport developments, which create local booms in land rents, mortgage interest tax deductions and other indirect housing subsidies, rent regulation, etc.). Wilensky concludes that A 'bewildering array of fiscal, monetary, and other policies that affect housing directly and indirectly – even remotely – have made the task of comparative analysis of public spending in this area nearly impossible' (Wilensky, 1975: 7).

Both of Wilensky's reasons for excluding housing are ultimately unconvincing. Concerning the poor quality of housing data, the preliminary data he presents in a footnote on housing and social security give a tantalising glimpse of what might have been possible had he persevered (see Schmidt, 1989a for a preliminary analysis along the lines abandoned by Wilensky). The argument of 'complexity' is half right, but for the wrong reasons because it applies, of course, to every dimension of social structure. There is nothing inherent in housing that makes it more *complex* than, say, health or education. They, too, are affected by a 'bewildering array of fiscal, monetary and other policies' that affect them 'directly and indirectly – even remotely'.

All the same, Wilensky is groping after an important feature of housing that ultimately eludes him, even though he is dimly aware of it. It is that housing manifests a high degree of 'embeddedness' in social structure. Its very pervasiveness in terms of influence on life styles, urban form, welfare, and patterns of household consumption make it at the same time central to understanding welfare yet conceptually elusive. It is this embeddedness that makes housing qualitatively different from, say, health or educational institutions. This is made clear if we consider how education, for example, could be radically reorganised by breaking up large schools into neighbourhood ones, or by combining schools with universities. The impact on social structure would be far less than, say, reorganising housing so that everyone lived in collectives, or high-rise flats, or multiple occupancy.

The special position of housing within welfare therefore causes considerable problems of analysis and it is therefore not surprising that Wilensky shies away from tackling it. Nor is Wilensky alone among welfare researchers in so doing. Housing is strikingly absent from comparative welfare research. It tends to be either omitted entirely or included as one item in a catalogue of separately treated welfare areas (see, for example, Esping-Andersen, 1985: 179–90).

The neglect of housing by comparative welfare researchers – and indeed by welfare research in general – is therefore, paradoxically, testimony to the importance of housing to welfare rather than its insignificance. Its lack of conceptualisation by housing researchers has been a deterrent to social welfare researchers addressing housing centrally. And, as we have seen, when housing welfare has been centrally addressed by researchers versed in both housing and general welfare (Donnison and Ungerson, 1982), it has been treated as an autonomous welfare area with its relationship to welfare in general remaining unproblematic.

Turning to the relationship of housing to theories of social change in welfare research, it is clear that housing must be a major constraint. The impact of housing on households' expenditure patterns over long periods and its effect on residence patterns, for example in terms of settlement densities, means that existing stocks of housing place considerable limits on the introduction of major welfare changes. Since housing is so deeply embedded in social structure and is so difficult to disentangle from other forms of welfare, it follows that housing tends to figure more in theories of change which deal with longer historical periods and with more enduring deep-seated changes, rather than with short-term or cyclical changes.

This no doubt explains why housing appear more prominently – though still only marginally – in evolutionary and divergent theories of welfare development than in cyclical and episodic theories. It may be thought to be particularly pertinent to labour movement theory because of the prominence given in it to ideology. Housing has always been a highly emotive subject, partly because shelter is such a basic and fundamental element in the social construction of life styles. Indeed, in industrial societies, to be homeless is a far more severe and alienating experience than any other – including unemployment – and can rarely be sustained for long without leading to severe personal and social strain, not to say collapse. But, over and above this, housing is emotive because of its associations with home-making: home as 'castle', 'hearth and home', 'nest-building', etc. (Gurney, 1990). This is what makes housing such an ideologically sensitive area of social life, and, as we shall see in Chapter 7, provides fertile ground for myth-making, often in a broader ideological context.

This makes it all the more surprising that labour movement theory and other theories which focus on ideology and class as key explanatory variables in social change have not placed housing more at the centre of analysis. But, once again, the reason for this must be sought in the limitations of the comparative housing research literature, which has tended to be heavily biased against ideological explanations of the deep structure of housing and its impact on society. Housing has been understood either in narrow party political ideological terms, particularly in traditional unexplicated policy analysis, or in terms of structures of provision to be understood in terms of uneven development or variations deriving from historical specificities, or even in terms of innate drives.

CONCLUSIONS

Developments in welfare research are likely to have a major impact on housing research over the coming decade. This is true in terms of both concepts of welfare and international comparative research into welfare. Developments in both of these areas are breaking ground which housing research continues to ignore to its detriment. A more theoretically grounded housing studies must take welfare issues and problems centrally into account. The key issue is to unravel the complex relationship between on the one hand, housing welfare and other forms of welfare and, on the other hand, housing and the wider social structure in which it is embedded. Ultimately, I believe that housing studies will be able to make major conceptual contributions to welfare research. For the purposes of this discussion, however, these conclusions constitute an important part of developing a divergence thesis in comparative housing research. That task is tackled in the third section of this book.

Part III

Toward a divergence thesis in comparative housing and research

6 Ideology and divergent social structures

INTRODUCTION

In Chapter 7 I shall argue that major structural cleavages, resulting from a process of divergence between industrialised societies of European origins, can be understood in terms of the emergence, over long periods of time, of collectivist or privatist social structures. Before looking at this, however, I will develop in Chapter 6, a framework for understanding the dynamics underlying the process of divergence at a more general level. In particular, I want to examine the relationship between ideology, social structure and the state.

My starting point is where I left off in Kemeny (1981), namely, in constellations of values that are associated with the social organisation of urban form, welfare, housing, the family, gender divisions of labour, etc. The cement that holds these together is ideology and ideation in general. I will argue that different kinds of industrial society are characterised by differently constituted overarching and (to some extent at least) internally consistent ideologies. To this end I adopt a modified form of the so-called 'dominant ideology thesis', normally associated with the need for a dominant ideology to legitimise capitalism (see Abercrombie *et al.*, 1980).

However, my usage of the concept differs from this somewhat rigid and limited formulation in a number of important respects. Crucially, I see ideology as being not secondary or derivative of social formations but central to the way in which social institutions are constituted, sustained, and changed. It is ideology which provides the motivation for action and which channels that action into the creation and perpetuation of social forms in its own image. A fundamental proposition of this thesis is therefore that ideology is not merely a reflection of major social formations but is interactive and plays an important – even decisive – role in determining the kind of society which develops. Further, I argue that there is not one dominant ideology for the legitimation of industrial societies but several possible dominant ideologies, the establishment of one of which results in a society developing in ways more or less consonant with the values expressed through it. In this way, to a greater or lesser extent, societies develop along different paths, thereby producing a process of divergence among societies which otherwise share a similar economic basis. Finally, I am interested in the dynamics that lie behind the way in which a dominant ideology develops, or fails to develop, in a society.

In this chapter, therefore, I want to examine the concept of dominant ideology and its application in the literature. I want to show how divergence can be understood in terms of the establishment of a dominant ideology and its associated mode of discourse, how these frame and delimit the major social, cultural, and political debates within a society, and how they act as guidelines for the social construction of institutions through interpersonal interaction.

To this end, the discussion comprises two parts, relating to the macro and micro processes respectively through which divergence takes place. This division underlies the approach to the agent–structure problematic that I use. It is not based on the programmatic theory of structuration, which provides no practical guidelines to integrating agent and structure in empirical analysis. Rather, it is based on the constructivist perspective in which interpersonal interaction possesses 'emergent properties' which lead both directly and indirectly to changes in larger scale patterns of interaction in both anticipated and unanticipated ways.

In the first part, after defining the concepts of ideology and myth, I consider macrostructural issues surrounding the concept of dominance and hegemony, particularly in relation to the state. I take as my starting point an evaluation of the work of Abercrombie *et al.* (1980) and Gramsci (1971) before defining my own usage of the concepts of hegemony and political dominance. I then show how this approach is more open to generalisation about social change than the more limited labour movement theory and how it can be used to understand processes of divergence between industrial societies under different conditions of hegemony.

In the second part I look more closely at how this is accomplished in the micro order of day to day interaction in terms of agency. Using negotiated order theory, I relate the idea of ideological dominance to the concept of a negotiated order. In particular, I look at the ways in which ideologies manifest in public policy in the form of powerful myths, and how these are sustained and changed in everyday interaction in political institutions.

MACROSOCIAL ASPECTS

The concepts of ideology and myth

The concept of ideology is one of the most problematic in the social sciences. For my purposes I do not intend to attempt a reformulation or even to engage in any exegeses of standard works. I content myself with presenting the usage that I adopt as a rough guideline to the discussion that follows.

By ideology I mean a loosely organised set of ideas defining human nature (the individual), the principles underlying the organisation of social life (civil society), and the values that govern the political order (state). Ideologies include postulated underlying principles, assumptions concerning the nature of society, and basic values governing behaviour.

I make no assumptions concerning the relationship of specific ideologies to specific social formations, groups, or classes, as, for example, in the Marxist concept of 'false consciousness' when a class adopts an ideology that is counter to its objective interests. Nor is ideology to be understood as some distorted version of an absolute truth, as is commonly conceived in the positivist tradition. Thus, for example, the system of scientific thought must be understood as itself being a specific ideology in the terms defined above (for a useful review of the concept of ideology as used by different social theorists see Larrain, 1979).

By myth I mean an application of the principles underlying a specific ideology to particular cultural and social circumstances that takes the form of a moral tale, or image, that is symbolically illustrative of the ideology. Myths are morally and emotionally charged descriptions of specific instances of social reality intended to concretise the abstract principles of an ideology. As de Neufville and Barton (1987) put it:

Myths are an important source of meaning, even in modern societies. They provide analogies which help make sense of events and provide simplifications of a more complex reality. Myths have a moral component, speaking of good and evil or simply providing object lessons. Because they are well known in a community they provide shared rationales to behave in common ways. Because they take dramatic form and touch on deeply held values, they evoke strong, shared emotions. They are created in a particular culture from its repertoire of images, symbols, characters and modes of actions.

(de Neufville and Barton, 1987: 182)

Myths must therefore be understood as concrete expressions of ideologies in everyday life. They includes images of, for example, the sturdy independent farmer, the stakhanovite, or the good samaritan, as well as moral tales such as the Robinson Crusoe story of overcoming adversity through individual effort.

Dominant ideologies and modes of discourse: Abercrombie, Hill and Turner

Abercrombie et al. (1980) provide a useful discussion and critique of what they term 'the dominant ideology thesis'. This refers to the strand of thought among Marxists which explains the continuing resilience of capitalism in the face of the growing crisis in terms of the development of a strong ideological 'cement' that ensures continued compliance by workers to the system. The authors argue against this type of explanation, preferring to explain continuity in terms of basic economic imperatives. Their work represents a particularly detailed and wide-ranging examination of the dominant ideology thesis and so provides a good starting point for this discussion.

Abercrombie *et al.* (1980) attempt to identify dominant ideologies in three European case studies: feudalism, early capitalism, and late capitalism. They argue that distinctions must be made between dominant ideologies held by ruling and subordinate classes and that different degrees of incorporation can be distinguished, largely dependent upon the effectiveness of the transmission of the dominant ideology from ruling to subordinate classes. They conclude that:

> In feudalism and early capitalism there are indeed identifiable dominant ideologies. These are held by the members of the dominant social classes but not by those in subordinate classes. One reason why lower classes do not hold these ideologies is the weakness of the mechanisms of transmission. We shall show that this state of affairs is rather different from that which obtains in late capitalism, where the dominant ideology is much less well defined, is made up of a number of disparate elements and contains several internal inconsistencies. In late capitalism there is some ideological incorporation of the working class, though less than has been recently believed, and the mechanisms of transmission are well developed and partly effective.'
>
> (Abercrombie *et al.*, 1980: 128)

They argue that elements of a modern dominant ideology include managerialism, accumulation, state neutrality and state welfare, respect for hierarchy and deference to authority (Abercrombie *et al.*, 1980: 137–8).

In a later study the authors examine the relationship between individualism and capitalism, but their emphasis has shifted from the *content* of dominant ideologies (and the extent to which these are accepted by different classes) to *modes of discourse*. They argue that 'Discourses may neither create nor sustain an economy, *but they do give it a certain shape by constituting the economic subject in a particular way*' (Abercrombie *et al.*, 1986: 184; emphasis in original). Here, then, the authors are more concerned with the existence of dominant modes of discourse than with dominant ideologies. They argue that, in the transition from early to late British capitalism, the significance of individualism as a mode of discourse has changed and become broken down into a number of discourses, none of which is necessarily dominant:

> individualism became appropriate to British capitalism in its early stages because it constructed the individual as the economic subject. As capitalism developed in the late nineteenth and twentieth centuries, individualism became less relevant, losing its position of dominance and leaving a plurality of discourses of the individual
>
> (Abercrombie *et al.*, 1986: 188)

Unfortunately, it is not clear from these two books how the concepts of ideology and of modes of discourse relate to one another. It would appear that the former refers to a set of inter-related beliefs, such as that provided by Mediaeval Catholicism or individualism in early capitalism, while the latter refers to a broader and more elusive *weltanshauung*, or possibly to a style of presentation, such as bureaucratic language. Be that as it may, Abercrombie, Hill and Turner appear to be arguing that modern British capitalism is characterised by a diversity of ideologies (Abercrombie *et al.*, 1980: 128) and modes of discourse (Abercrombie *et al.*, 1986: 188), none of which is necessarily dominant. This in turn leads the authors to suggest that the diversity opens up the possibility that *collective* modes of discourse can become dominant under certain conditions:

> None of this holds out any certainties for the future, it is perfectly possible for a discourse emphasising collective rights to attain some social dominance and to define a collective economic subject for capitalism.
>
> (Abercrombie *et al.*, 1986: 188)

This in its turn implies – though it is nowhere explicated – that a process of divergence between societies with individual as against collective based modes of discourse is possible, and that modes of discourse can provide the means for the emergence of different kinds of social structure.

The work of Abercrombie and his colleagues is by far the most thorough and challenging attempt in the recent literature to understand the role of ideology and discourse in modern societies in relation to the buttressing of an economic and social order. But their treatment is essentially descriptive and correlative. They are mainly concerned to show that in different historical periods different ideologies and modes of discourse tend to be characterised by different degrees of dominance as a result of varying degrees of incorporation and successful transmission. They do not attempt to show the dynamics whereby one ideology or mode of discourse succeeds in becoming dominant at the expense of others. Nor do they attempt to define 'dominance'. Indeed, one must conclude that the difference between whatever the authors mean by an ideology and a mode of discourse implies that the dominance of one is likely to take a different form from the dominance of the other.

For example, for an ideology to be dominant, according to Abercrombie *et al.* (1980), it would seem that concrete sets of beliefs must be widely diffused both within and between classes. However, for a mode of discourse to be dominant (Abercrombie *et al.*, 1986), it is not at all clear that incorporation is necessary. It may be sufficient merely for *public* discourse (for example in politics and culture) to be conducted in a particular mode for it to become dominant. Thus, for example, it may be that class modes of discourse remain distinct from one another but that one becomes dominant in key public areas, such as politics and the media, the result being that members of a subordinate class continue to use their class mode of discourse in intra-class situations but feel compelled to adopt, to their disadvantage, the dominant mode of discourse in public.

This distinction between a mode of discourse as a form for, or a vehicle for the conveyance of, ideological content is the most likely interpretation of the usage of Abercrombie and his colleagues and is the one which will be used here. However, it is important to be clear that the form–content distinction that this implies is often untenable or at best limited in usefulness, at least in empirical analysis. It is not always possible to separate the content of ideology from the discourse which is transmitting it.

Finally, the authors do not have very much to say about the interaction between ideology and social structure. It is therefore unclear whether the existence of a dominant ideology has any implications for the way in which society is organised. The evidence for this must be inferred from their conclusions that the existence of a strong and widespread dominant ideology is not a necessary precondition for a stable capitalist society. In the final analysis, then, Abercrombie, Hill and Turner would appear to adopt the position that the existence or not of a strong dominant ideology does not result in significant social structural variations.

Given the limitations of the formulation of the dominant ideology thesis and its descriptive character, it is necessary to look elsewhere to obtain a more dynamic and constructivist understanding of the nature of ideologies and the establishment of dominance, and the relation of ideology to social structure. A useful approach is that provided by Gramsci in his concepts of hegemony and political dominance.

Hegemony and political dominance: Gramsci

The work of Gramsci had been neglected in anglophonic research until the late 1970s when there was a spate of exegeses and applications (Bloomfield, 1977; Laclau, 1977; Laclau and Mouffe, 1985; Salamini, 1981; Sassoon, 1980, 1982; Simon, 1982; Urry, 1981). Even today, Gramsci is considerably under-rated as a Marxist theorist, partly because his complete works have still not been translated into English, but more because of the incoherent, repetitive, coded, and sometimes contradictory nature of the prison notes, written under almost impossible conditions, that comprise such a major part of his voluminous output (Gramsci, 1971). His work is particularly neglected in political science and urban sociology where Althusserian structuralism has come to dominate Marxist theory, largely through the work of Poulantzas and Castells, respectively. Just how neglected Gramsci's work is can be appreciated when it is realised how little his concept of hegemony figures in the work on dominant ideologies and modes of discourse discussed above. Yet it is arguable that Gramsci's development of a theoretical framework applicable to modern complex industrial society makes him the most important Marxist theorist since Marx.

It is important to realise that Gramsci was writing at a time when the success of the Bolshevik revolution overshadowed all socialist thought and political strategy. This success had strengthened the position of those socialists who believed that their prime task was to achieve revolution through control of the state, a belief that had already become firmly entrenched in socialist thought following the defeat of the libertarian wing of the socialist movement at the First International. As a West European Marxist, however, Gramsci was very aware of the differences in the social conditions pertaining there to the situation in Russia at the time of the revolution. In particular, he was struck by the greater complexity and resilience of the social structure of Western European societies and the commensurately lower prominence of the state and political institutions (Gramsci, 1971: 238). This led Gramsci to argue that, while the strategy of the bolsheviks in Russia to sieze the state apparatus was sound precisely because the state was so powerful in relation to a poorly-developed civil society, a similar approach in Western European countries would be doomed to failure. In societies with complex social institutions and dense networks of relationships outside of politics, an essential prerequisite for revolution was to gain a leading position in relation to civil society. Gramsci therefore distinguished between, on the one hand, the exercise of leadership by a class whereby it attains a position of moral and ideological leadership – hegemony – in the wider society and, on the other hand, the exercise of domination through the control of political institutions:

> What we can do for the moment is fix the two major superstructural 'levels': the one that can be called 'civil society', that is the ensemble of organisms commonly called 'private', and that of 'political society' or 'the State'. These two levels correspond on the one hand to the function of 'hegemony' which the dominant group exercises throughout society and on the other hand to that of 'direct domination' or command exercised through the State and 'judicial' government.
>
> (Gramsci, 1971: 12)

For Gramsci, hegemony (leadership) precedes domination and is a precondition for it, since if ever hegemony were lost then domination would be difficult to maintain:

> A social group can, and indeed must, already exercise 'leadership' before winning governmental power (this is indeed one of the principal conditions for the winning of such power); it subsequently becomes dominant when it exercises power, but even if it holds it firmly in its grasp, it must continue to 'lead' as well.
>
> (Gramsci, 1971: 57–8)

Gramsci therefore placed great emphasis upon the development of moral, intellectual, cultural, and social leadership (hegemony) in order for a social group to gain – and sustain – dominance in a society. He argued that intellectuals do not constitute a class, but that each class produces its own intellectuals who play a leading role in the formulation and articulation of class ideologies (Gramsci, 1971: 5–23). An essential element of attaining hegemony is the building of social and political alliances, or what Gramsci tends to call a 'bloc'. By this he normally means an alliance of groups under the leadership of one group and the assemblage of a hegemony based on integrating movements and ideologies that transcend class interests, including such movements as nationalism and minority rights (and today we might add, for example, environmental, peace, racial equality, and women's rights movements), in order to create national leadership.[1]

The attainment of leadership and ultimately dominance is not for Gramsci an inevitable progression resulting from the development of the logic of capitalism. Rather it is something to be striven after, both to attain and to maintain once gained. Social groups therefore constantly struggle, first to weaken the existing hegemony of another group, then to attain hegemony, and then to hold and strengthen it. It follows that groups may lose and regain hegemony as first one group then another gains and then loses ascendancy in civil society.

One example of this is what Gramsci terms, somewhat misleadingly, 'passive revolution' (Gramsci, 1971: 59), whereby the bourgeoisie succeeds in reimposing its slipping hegemony, thereby forestalling active revolution. He defined the Italian *Risorgimento* as an example of such passive revolution brought about by bourgeois forces, in this case in alliance with an outside state (Piedmont). 'Bonapartism' was another such example, whereby the slipping hegemony of a bourgeoisie may be buttressed, the prime contemporary Italian example being Mussolini. In any event, it is clear that for Gramsci hegemony is a state of affairs that must be constantly maintained and defended against challenge by other groups.

Finally, in this brief overview of Gramsci's work on the dynamics of hegemony we may note that hegemony is understood as leadership in its broadest possible sense. When Gramsci died, he had commenced a large programme of research on a wide range of cultural expressions of hegemony, including literature and drama in which he was particularly interested. And while Gramsci himself did not provide a full definition of the term 'hegemony', the following definition gives a good indication of the breadth of the term in the sense used by Gramsci:

> By 'hegemony' Gramsci seems to mean a sociopolitical situation, in his terminology a 'movement' in which the philosophy and practice of a society fuse or are in equilibrium; an order in which a certain way of life and thought is dominant, in which one concept of reality is diffused throughout society in all its institutional and private manifestations, informing with its spirit all taste, morality, customs, religious and political principles, and all social relations, particularly in their intellectual and moral connotation. An element of direction and control, not necessarily conscious, is implied.
>
> (Williams, 1960: 587; cited in Sassoon, 1982: 95)

The relevance of Gramsci as one starting point for the development of a divergence thesis, along the lines proposed in this book, may be summarised as follows:

1 He provides a Marxist account of social change which while theoretically informed, is non-deterministic.
2 He gives primacy to superstructural factors and, in particular places, great emphasis to what he terms 'civil society' which is a useful way of distinguishing between state and society (see Urry, 1981 for a discussion of this).

3 His concept of ideology is based on relativist principles both in the sense that he does not define ideology as a distortion of some absolute truth, as Marx does (Larrain, 1983), and in the sense that ideology as a set of beliefs held by a class is not something derivable from class position but is socially constructed in the process of attaining hegemony.

4 The state and political institutions are not seen as a passive 'black box' through which social groups implement their interests. On the contrary, there is a complex dynamic between leadership of civil society and domination of political institutions.

5 Gramsci's approach has many affinities with the hermeneutic tradition in sociology, and particularly with constructivist sociology. He places great emphasis on the ability of groups to develop hegemony through a process of social construction. This is an important difference between divergent and unilinear accounts of social change, and constitutes the basis for the divergence thesis developed below.

Hegemony and dominance: a reformulation

Despite its general relevance, the work of Gramsci is problematic and underdeveloped and, above all, contradictory and open to many different interpretations. Gramsci's work therefore constitutes only a starting point for developing the concept of hegemony and political dominance. In this section I attempt to spell out some important dimensions of hegemony and political dominance that make the concepts more applicable to the analysis of changing social structures and divergence in particular.

The concept of hegemony is best approached in terms of an ideal type. We may therefore say that a group, in its ideal typical form, attains hegemony when its leadership is established across the whole gamut of social life. In practice, however, it is not necessary, nor even likely, for hegemony to exist in such a pure or extreme form. It is conceivable, indeed probable, that a group may establish hegemony by exercising leadership in a number of key areas such as intellect, in terms of political principles, and morality while leaving disputed less central areas such as art forms and educational principles. Such 'partial hegemony', as long as it is sustained in key areas of civil society, may be sufficient for overall hegemony to be attained, which in turn can form the basis for political dominance. Partial leadership must be seen as the rule rather than the exception, and as providing the conditions to enable an existing hegemony to be challenged and ultimately displaced.

There are good sociological reasons for such a distinction. Hegemony means literally 'leadership'. But leadership can take many forms and can be more or less complete or it can be partial. This leaves open the possibility for a group to exercise leadership in a number of key areas, but by no means all areas, and still exercise hegemony in relation to society as a whole. Hegemony may therefore be exercised under different conditions of leadership, ranging from the more unusual situation in which leadership is fully comprehensive to the more usual case in which hegemony is constructed on the basis of leadership in a few key areas. This in turn means that the degree to which hegemony is securely held may vary considerably.

Much the same may be said for the concept of domination. Ideal typically, political domination may be said to exist when a group that exercises leadership of key areas of civil society and has thereby attained hegemony attains control of the state. But the state is a complex and diverse set of institutions, including judicial and repressive apparatuses. The complexity of this is well recognised in particular by Althusserian Marxists (for a somewhat daunting taxonomy see Clark and Dear, 1984). It is unlikely that any group could establish total domination over all political institutions and hold it over a sufficiently long period to enable its programme to be implemented. Nor is this necessary: it is normally sufficient for domination to be maintained by control of key elements of the state apparatus over reasonably long, though not necessarily unbroken, periods of time.

Dominant ideologies and divergence

The relevance of the dominant ideology thesis lies in this discussion not with its applicability to explaining the persistence of capitalism as a system, but with its usefulness as a means of understanding why large differences emerge between industrialised and urbanised societies of European origin. We have seen earlier in this chapter how Abercrombie *et al.* (1986: 188) imply that different modes of discourse can emerge, and that they concede the possible emergence of collectivist modes of discourse. My argument takes this line of reasoning further by arguing that differences between basically similar industrial societies must be understood in terms of *the emergence of different kinds of dominant ideologies*, reflecting the success of certain vested interests in defining – over decades, or more usually generations – the parameters of what is and what is not possible, efficient, desirable, etc. Divergence can therefore be understood in terms of the emergence of different dominant ideologies and the way in which these have shaped basic social institutions over extended periods of time. That is, the emergence of a dominant ideology in a society is not merely an ideational change, but has far reaching implications for the way in which society is organised, since the dominant ideology informs social action and leads to changes in social structure which in turn mirror, or more accurately refract, the aims of the ideology.

The relationship between ideology and social structure, mediated by the existence of hegemony and reinforced by the exercise of state powers through political dominance, is vital for understanding why societies develop characteristic social structures. Ideologies both reflect the organisation of social structure and provide guidelines for the direction in which it is to be changed. The establishment of hegemony therefore provides the social conditions in which it is possible to change, at least in some respects, social structure so that it reflects the ideals embodied in the ideology. Political dominance is a means of reinforcing and entrenching a dominant ideology in a social structure by using the state to form basic laws and to encourage forms of institution that are consonant with the dominant ideology and to disadvantage those which are not. Political dominance must therefore be understood as an important complement to hegemony, in that it legitimises the ground assumptions of the dominant ideology and provides the politico-legal framework and the subsidies and fiscal penalties in which social structure can be made to conform more closely to the tenets of the dominant ideology.

The dominant ideology thesis I want to propose here can be seen in some respects as a development of labour movement theory, whereby the extent to which the welfare state is developed can be understood in terms of the strength of the labour movement and its ability to wrest concessions from its opponents and to form social policy after its own interests (Offe, 1972; Hicks *et al.*, 1978; Esping-Andersen *et al.*, 1976; Korpi, 1978, 1983). The great strength of labour movement theory is that it embodies such a link between ideology and the state in the very nature of the approach.

But labour movement theory has a number of major weaknesses. The first is that it fails to problematise the link between the labour movement and the outcome. It simply assumes that there is a 'natural' vested interest that is unquestioned, even if in practice it must be diluted by practical political compromises. Labour movement theory also tends to focus disproportionately on political action, and on the control of the state in order to push through social security programmes. It thereby tends to neglect, though it does not ignore, broader social struggles and forms of action. A further limitation of labour movement theory is that it focuses only on one class interest and thereby neglects other dimensions and other ideological formulations than those of the labour movement. Labour movement theory provides no frame of reference for understanding bourgeois, middle class or agrarian movements, for example. Labour movement theory also fails in general to understand labour movement gains in terms of the establishment of a dominant ideology, or in terms of the broader context of countering hegemony and establishing a new hegemony. Finally, labour movement theory has paid almost no attention to the interaction between ideology and social structure. Almost all work has instead focused upon the welfare state and social security programmes. But, crucially, there has been a serious neglect of the importance of ideology for the very manner in which social institutions are formed and organised.

Going beyond labour movement theory formulations, therefore, it is fruitful to define a dominant ideology as an ideology which has succeeded in becoming widely accepted in a society, to the extent that the social, economic and political agenda-setting is largely determined by the parameters of that ideology's own mode of discourse. In some cases such a dominant ideology will be consonant with the collectivist values of a labour movement, but in others it could well be an individualistic and anti-welfare ideology such as is found in Australia or the USA, reflecting perhaps employer interests or a powerful conservative/bourgeois political party. In yet others it could be the co-operative mutual aid ideology of a syndicalist movement. Of course, in some societies, and under some conditions, a dominant ideology may not succeed in becoming established at all. In others, there may be two or more sectional ideologies competing for dominance, with none actually succeeding, or with resulting structures being formed in terms of corporatism. Indeed, it may be possible to define corporatism as the product of competing sectional ideologies under conditions in which none can gain ascendancy over the others. In yet other circumstances an ideology may be dominant yet not sufficiently so to create the conditions for hegemony.

Further, the dominant ideology at any one time in a society may be a composite or compromise (for example centrist), such as 'Butskellism' in Britain during the early post war period. It may also be an unstable compromise with certain elements seeming to gain ascendancy at one time over others. Thatcherism can be seen as an example of this, which, if successfully sustained over several decades, would lead to a major reformulation of the dominant ideology, with far-reaching repercussions for the nature of society. A similar attempt to redefine basic values, albeit in its very early stages, can be seen in the collapse of communism in Eastern Europe.

This points to the potential importance of the role of the state in the process of establishing a dominant ideology, or challenging an already established one. I intend to argue in the next section that the state and public policy can and must be seen not simply as a passive medium for vested social interests but as a key set of institutions for the determination of agenda-setting, and for reinforcing hegemony by introducing changes in social structure through changes in laws and directing taxation/subsidies.

But dominant ideologies need not take such explicit political expression as the above examples imply. Indeed, the hallmark of a really powerful dominant ideology is that it does not take controversial and sectarian political expression simply because it is so deeply entrenched socially and so ingrained in public modes of discourse that the political agenda remains 'hidden': implicit and taken-for-granted, with most political parties accepting it without question and with dissent limited to marginal issues or restricted to marginalised groups. Dominant ideologies may therefore be of different kinds, with differing degrees of stability. Some may be almost hegemonic in nature. Others may be constantly under challenge. Yet others may be compromises and accommodations in a finely balanced and continual conflict over defining the boundaries of the possible. Yet others may reflect deep corporate cleavages in society, over race, religion, or class.

The emergence of dominant ideologies in societies – whether collectivist or privatist, or whether of a more indeterminate or possibly a compromise character – helps us to understand why societies with an otherwise basically similar economic base can develop, over the decades and generations, radically different social structures. The importance of a dominant ideology lies in the ways in which its principles form the basis for social organisation, in terms of both changing that of existing institutions and, as social change creates the conditions in which new forms of social organisation are needed, determining the form taken by new institutions. It thereby helps us to understand why societies can apparently set out on a path of social change and development which is very different from otherwise similar societies.

The long-term instability of many dominant ideologies also helps us to explain why, during certain periods of history, certain societies appear to become more similar to one another (converging) while in other periods this is not the case, and at yet other times they may be diverging. One reason for this is to be found in the international 'knock-on' effect that dominant ideologies can have. The obvious recent example of this is the way in which the previously discredited belief in the free market has been resurrected and widely embraced almost irrespective of political affiliation.

It is worth while at this point to say something more about the relevance of the concept of hegemony for corporatist theories of the state. I have already suggested that corporatism can be understood in terms of the inability of one ideology to attain dominance and where several competing ideologies exist. But hegemony can take many forms. Less than total forms of hegemony can, under certain conditions, result in a special kind of corporatism when strategic alliance-building embraces most of the important interests in a society. Opposing interests may then be incorporated into administrative and political decision-making, in what might be termed 'hegemonic corporatism": that is, corporatism under the leadership of a strategic alliance (Sweden is an example of this: see:

Chapter 8). It may therefore be possible to conceive of different forms of corporatism depending on the existence of competing or dominant ideologies.

This may be illustrated by means of a simple model (Figure 6.1), distinguishing between dominant and fragmented ideologies on the one hand and political power, which is either inclusive or exclusive of opposition elements, on the other.

Figure 6.1 Ideology, exclusion, and political power

Ideology	*Political power*	
	Collaborative	Competitive
Dominant	Hegemonic corporatism	Political dominance
Fragmented	Corporatism	Class pluralism

The most fractured is found when no ideology is dominant and political power is held competitively by one or another of the main social groups for short periods in unstable succession. This is the classic pluralist or class model of social power, characterised by the situation in many anglo-saxon countries. Ideological fragmentation where a high degree of collaboration has been attained produces competitive corporatism, in which compromise characterises the accommodation of the primary ideological groupings that constitutes classic European corporatism.

Where there is a dominant ideology and political power is held competitively, the strategic alliance that has attained hegemony in civil society will hold political power to the exclusion of other interests in society, reflecting classical political dominance. Thatcherism may be such a hegemony in the making. But where a dominant ideology is so powerful that it includes all major social interests, a form of corporatism can result, which is not between more or less equal parties but which is led by the hegemonic social group, and I will suggest in Chapter 8 that Sweden is such a case.

This, then, constitutes the outline of an approach based on the study of competing ideologies which lead, under certain circumstances, to the emergence of one ideology as a dominant ideology, leading in turn to the establishment of hegemony and ultimately to domination through political structures, and resulting in the long run in divergence from societies in which a different dominant ideology has emerged. But so far nothing has been said about the process by which a dominant ideology is created, sustained, and changed. To understand this process it is necessary to make a brief excursion into the agent–structure problematic and the hermeneutic tradition in sociology, touched upon in another context in Chapter 1.

MICROSOCIAL ASPECTS

Negotiated order theory

In Chapter 1 we saw how work on the agent–structure problematic in sociology came to the attention of urban and housing researchers through the work of Giddens and his interest in time and space, and how, because of his dismissal of the symbolic interactionist or, as it is often known today, the constructivist perspective, it has been neglected. However, this perspective is peculiarly suitable to understanding the processes through which ideologies are generated and hegemony is attained because it is based on the premise that society is the product of definitions of reality held by people, and that such definitions are sustained and changed through interpersonal interaction to become the basis for social action which in turn generates the social organisation that frames and limits future definition changes (Berger and Luckmann, 1966; Blumer, 1969; Mead, 1934).

In the constructivist perspective, then, social structure is not something that exists *deus ex machina*, and that stands over individuals as a totally unresponsive structural constraint. Rather it is generated and modified in and through the process of interpersonal interaction. Clearly, individuals are constrained by structures – and often severely so – but only to the extent that they accept definitions of reality which others hold and which they then take into account in deciding their own lines of action.

Individuals construct social reality through their everyday interactions with others, in which, with very different resources available to them, and in relation to established practice, they struggle to impose their own beliefs, values, definitions, etc. on the significant others with whom they interact (Berger and Luckmann, 1966). In so doing, they collectively create, change, and sustain group-reinforced meanings and understandings which in turn are seen as, interpreted as, and believed to be, structural constraints on future actions.

This is simply illustrated in terms of personal interaction in Berger and Kellner's (1968) analysis of the social construction of marriage. They show how marriage – like all relationships – is sustained and changed by a constant stream of talk. This takes place in everyday conversation, day in and day out, in which partners modify their individual definitions of themselves and each other in numerous minor and major ways. By this means the partners eventually arrive at mutually agreed definitions of themselves in relation to significant others (in-laws, friends, etc.) which are often radically different from those held prior to marriage and which are themselves subject to constant re-evaluation (for example, defining out previous boozing companions, or alternatively integrating them into the relationship in a new way). These definitions then constitute the basis for social action and the way in which the couple organise their lives. This process of redefinition is constant and continues throughout the marriage, as new individuals impinge on the relationship and as the circumstances of existing individuals change over time.

On a more macro scale, the concept of a 'negotiated order' has been developed (Day and Day, 1977; Fine, 1984; Hall and Spencer-Hall, 1982; Maines, 1978, 1982; Martin, 1976; O'Toole and O'Toole, 1981; Scanzoni, 1979; Strauss *et al.*, 1963, 1964; Strauss, 1978, 1982) to explain how organisations and other meso-level social institutions develop their own characteristic networks of understandings, values, agreements, compromises, etc. These may change as old personnel leave and new personnel join and as outside circumstances – which are themselves the product of negotiated orders produced by interaction in other social institutions – change. As with micro-level social construction this is accomplished through the constant daily stream of talk.

This perspective has been clearly developed in the work of Collins (1975, 1981, 1987a). His view builds upon the idea that individuals form links in what he terms 'interaction ritual chains' that comprise networks of interaction. These networks provide pathways for the exertion of power and influence which have macro consequences. The concept of emotional energy and its importance in the social construction of reality has been developed by Collins (1984, 1987a, 1989). This can be illustrated by one of Collins' empirical applications in which he tries to explain the rise of schools of thought and paradigms in science and philosophy in terms of the creation of networks of interaction, critique, and collaboration between prominent scholars and their disciples. He argues that the major schools of thought are characterised by a surge of emotional energy generated by and within networks of interaction, stimulating new ideas and creating enthusiasm and a degree of intellectual solidarity and support which contribute towards new paradigm development. This is an important contribution to explaining social change, and one that has tended to be neglected in the literature, perhaps because it is seen as being too psychologistic in nature. Collins argues that interaction produces emotional energy – a sort of emotional charge – which both acts as the catalyst to further interaction and loads decisions and actions with meaning. This approach is potentially very fruitful for bridging the micro–macro link, and has been the basis of some of my earlier work (Kemeny, 1976a, 1976b, 1983b).[2]

In passing, it may be noted that the agent–structure dynamic implied by the perspective comprises a set of propositions that goes much further and is more practically useful in analysis than the concept of structuration. Giddens' dismissal of the constructivist perspective is made all the more curious because he uses the concept of 'locale' in the same sense as Strauss *et al.* (1963). More important for the purposes of this discussion is the centrality of definitions and the complex interplay between definitions of reality and lines of action which are pursued by individuals in co-operation and competition in order to implement definitions. It is this process that lies at the heart of how hegemony is established and maintained.

A good example of this can be seen in the acceptance within the international socialist movement of a statist strategy; a strategy of constructing (state) political dominance rather than (civil society) hegemony. This channelled energies into building political movements and so neglecting the social construction of grass-roots organisations for the collectivisation of social structure. The ideology determined the social forms which the class struggle took, and thereby gave rise to communism on the one hand and state welfarism on the other, while syndicalism and co-operation were neglected except in a few countries, particularly in Latin Europe.

In constructivist terms, then, an ideology may be said to be dominant when definitions of reality associated with a particular ideology are widely accepted. Hegemony refers to the ability of the members of a social group to impose their definition of reality upon the other members of a society, such that their definition constitutes the taken-for-granted assumptions that underlie everyday action, and, most importantly, informs the manner in which social life is organised. Political domination backs this up with the power of political institutions, by providing ground rules and financial and institutional support for the forms of social organisation that most closely reflect the definitions of reality embodied as central tenets of the dominant ideology. But ideologies are not monolithic systems of thought. They are only very general orientations to reality, often incorporating contradictory elements, and in order for these to take practical and concrete expression they need to be applied to specific forms of problem solving. These specific applications of ideology may be termed 'myths', since a myth comprises an interweaving of general principles of an ideology with a set of practical conditions that need to be dealt with.[3]

In the discussion that follows I am concerned with the place of myth in housing policy as a case study in public policy, which is intended to illustrate the practical working of domination arising out of hegemony.

Myth and the social construction of public policy

The concept of negotiation and of negotiated orders provides a useful framework for understanding myths. Myths do not suddenly manifest, full blown, nor are they free floating and randomly generated justifications siezed upon by policy-makers (whether public or private) on the spur of the moment. Rather, they derive from deep cultural roots and prejudices and are formulated in a process of interpersonal negotiation more or less systematically out of the material and social circumstances within which the interaction is set. Myths must therefore be understood as the products of interaction within specific cultural and social milieux.

The key ingredient in the social construction of myths is the stoking of emotional energy behind the myth. It is this which provides the myth with its potency or, more accurately, converts a neutral story into a moral tale. The whipping up of emotional support for a myth preparatory to the implementation of policy measures allows policy-makers to influence the moral limits within which policy debates take place. Myths behind alternative policies lose potency and the extent of their appeal declines. As a result their proponents are forced to conduct the political debates within the terms set by the proponents of the dominant myth.

Myths behind specific public policies gain in potency if they can be identified with larger myths in the wider society, and if they can be seen to be in harmony with the myths behind policies in other fields. This both creates the impression of consistency and has an amplifying and reinforcing effect, by locating policy solutions within a larger framework and by showing how policy measures in one field complement, and are in harmony with, policy measures in related fields. The emotional appeal of a myth in one area is considerably weakened if it is experienced as contradicting other, already well established, myths in other areas. It is for this reason that dominant ideologies are more successful when they are based on simple and unified principles of behaviour and morality (such as individualism, or collective responsibility).

These general and somewhat abstract observations constitute the raw elements which may be applied to an understanding of specific policies under identifiable historical circumstances. They may be summarised as follows. Myths must be understood as an essential component of the process of reality construction; a process of interpersonal negotiation which takes place continuously during interaction and which generates negotiated orders, comprised of values, beliefs, norms, standards, compromises, agreements, etc., which constitute the web of mutual understandings of social life in a particular institutional context. Myths are the more effective the greater the emotional energy that can be mobilised behind them. They are the more powerful the more they can be identified as belonging to wider and more potent dominant societal myths, and the more they can be seen as being consistent with the myths behind the policies in related areas. Since the potency of a myth is directly related to the emotional power that can be mobilised behind it, successful myths are almost always highly selective and one-sided interpretations of reality.

So far, the discussion has been limited to the idea of myths in a very broad sense. Use of the word 'policy' has been deliberately kept general to cover policy in a company, school, hospital or other organisation, or indeed any social institution with a recognisable power structure and hierarchy. However, I am particularly concerned with public policy because of the special nature of the state and its importance in buttressing hegemony.

The milieux in which public policy is socially constructed are first and foremost those which are constituted within political institutions: at the hustings, within political parties, in cabinets and ministries, among civil servants, and in the debating chambers of the legislature. There, the endlessly reiterated arguments and debates, conducted daily over many years and consolidated and reinforced in the media, are the seed-beds for the emergence, honing, and elaboration of the myths that underlie public policy-making. An early symbolic interactionist interpretation of politics in these terms is that by Edelman (1971, 1977), who examines political language in terms of metaphors, ritual, and the manipulation of beliefs. He argues that politicians are never able to deliver on promises because of the structure of political institutions and the unrealistic demands on their performance. Political institutions therefore generate a characteristic mode of operation based on the manipulation of ideas and images.

The insight that those who control political institutions have quite different priorities from specific vested interests and have, as we have seen Skocpol argue in Chapter 3, special responsibility for maintaining internal order and defence is supported by the work of Heclo and Wildavsky (1981). In a detailed observational study of policy-making by top British civil servants and ministers, they stress the importance of 'community' in framing internal political debates and thereby limiting demands made by outside interest groups. The place of metaphor and myth in framing debates among policy-makers is, unfortunately, not brought out by Heclo and Wildavsky, though it can be implied, especially from Wildavsky's work on culture (Douglas and Wildavsky, 1982; Wildavsky, 1987).

In recent years, however, there has been growing recognition of the importance of emotional appeal in the framing of policy across a wide spectrum of issues. It has been argued, for example, that penal policy is as much steered by 'irrational' feelings of outrage and revenge as by attempts at the rational control of crime (Garland, 1990). In an explicit study of political rhetoric in the case of green belt policy in planning, Rydin and Myerson (1989) show how certain fundamental, rhetorical and analogous themes can be identified, or what the authors, citing Burke, term a 'collective poem' in which the green belt is to the city as the garden is to the home.

More directly relevant still, in a recent article, de Neufville and Barton (1987) examine the central place of myths in the definition of policy problems. They argue that myths – in the sense of culturally shared and morally charged meanings expressed in story form – comprise a vital component in the construction of social policy and in providing the emotional energy necessary for mobilising sufficient support for its implementation.

One of the two case studies they use to illustrate the importance of myth in policy is what they term 'home ownership and the Jefferson ideal' (de Neufville and Barton, 1987: 185). They show how the myth of the yeoman farmer with virtues of self-sufficiency, hard work, and moral respectability formed an important image on which to build US policy towards the support and encouragement of home ownership. They trace the way in which the yeoman farmer image became adapted and modified to fit changing circumstance, particularly during the emergence of suburban home ownership under very different conditions from those which prevailed when the yeoman ideal originated. They show how the myth becomes even more attenuated in adapting it to the rise of the condominium (de Neufville and Barton, 1987: 192–3).

De Neufville and Barton make a plea for policy analysis to study the role of myths in policy-making, rather than either failing to acknowledge their existence or becoming an advocate of one or another particular myth as being 'a good thing' in pursuit of specific social goals. What they fail to do is to seek an understanding of myths in terms of wider ideological factors and as part of longer-term social change. One way out of this restricted policy-analysis approach is to integrate the concept of hegemony with myth-building in public policy, based on a constructivist analysis which relates the societal scale processes of change to micro and meso processes of the negotiated order.

CONCLUSIONS

This chapter has outlined the conceptual framework that I propose to use to develop a divergence thesis for understanding differences between modern industrialised societies. The framework comprises four major elements.

The first is the concept of hegemony and political dominance and the ways in which classes are able to build broad-based blocs to entrench hegemony or to challenge and break down the existing hegemony of other classes, and the importance of the use of the state to achieve this.

The second is the interactive relationship between ideology and social structure. Ideologies are not merely abstract systems of thought, but both reflect and can be used to transform existing social relations, institutions, and organisations. The displacement of one dominant ideology by another therefore brings in its wake changes in social structure that are consonant with the new dominant ideology.

The third is the socially constructed nature of hegemony through interpersonal negotiation and the social construction of dominant definitions of reality. It is the complex interplay between negotiated orders, constructed in different contexts, and the intended and unintended consequences of these for future interpersonal interaction that constitutes the link between individual action and societal outcome.

The fourth is the role of the state in the establishment, strengthening and challenging of hegemony. There are two components here: the ability of an historic bloc to use the state as a means of strengthening hegemony and the impact of the state itself over and above sectional interests. Underlying both of these is the concept of 'myth-building' which is particularly well illustrated in housing policy.

The relationship between ideology and social structure is crucial to the approach. A dominant ideology will in the long run produce changes in social structure that are consonant with it. The longer the ideology is dominant the greater will be the extent to which social structures begin to reflect the principles that the ideology embodies. And the longer the ideology is dominant the deeper into social structure the changes are likely to reach and the more profound and far-reaching they are likely to be. Political dominance, through control of law-making and subsidy directing, provides an important means of entrenching an ideology in social structure.

This conceptual framework can be used to understand the reason for major differences between industrialised societies along a number of different dimensions. It could, for example, be used to explain the differences found by Esping-Andersen (1990) between the three major types of welfare or the income-based welfare strategy of the Australian labor movement, both of which were briefly outlined in the previous chapter. It is, therefore, a general framework that is applicable to a wide range of differences between societies. For my purposes, I propose to use it to examine differences between industrialised societies in terms of privatist and collectivist ideologies and the ways in which these influence social structure. The next chapter outlines the main characteristics of this distinction. I argue that it constitutes perhaps the single most important cleavage between modern industrial societies, and that housing – or, more accurately, residence – comprises the key dimension by which it can be understood.

7 Divergent social structures and residence

INTRODUCTION

In this chapter I begin the task of developing a divergence thesis of social change among industrialised societies of European origins. The purpose of this chapter is to present the argument and some evidence for a divergence thesis in which residence plays a central part.

I start with a critical evaluation of my earlier comparative work. The first section may therefore be seen as a continuation of the critical overview of the comparative housing literature presented in Chapter 4. My own work suffers from rather similar general defects as those of the rest of the literature, at least in terms of lack of theoretical explication if not in terms of implicit unilinearity. Following this, I develop the idea of collectivist and privatist social structures and the relationship between these and forms of state welfare as a specific application of the dominant ideology thesis presented in the preceding chapter. I then review the evidence for the divergence thesis in the light of this.

LATENT DIVERGENCE IN *THE MYTH OF HOME OWNERSHIP*

The basic thesis in *The Myth of Home Ownership* (Kemeny, 1981) and developed elsewhere (notably Kemeny, 1978, 1980) was that there are wide and systematic disparities between the social structures of highly industrialised and urbanised societies of European origins that must be understood in ideological terms. The intention was to get away from broad generalisations about how such societies are similar, or about how the differences are mere 'variations'. Unfortunately, because of the undertheorised nature of my argument the thesis remained poorly explicated. This is clearly one reason why the argument has tended to be given much less attention than the concrete housing tenure issues which were intended primarily as the substantive - as distinct from the theoretical - focus.

For this reason, I am concerned here less with the discussion on housing tenures, which tends to be the part of my argument upon which most housing commentators have focused (Harloe and Martens, 1984; Hayward, 1986; Ruonavaara, 1987), and more with the part concerning comparative social structures. This is based primarily on Sweden, Britain, and Australia, as representing different forms of industrial society.

My basic argument was that housing cannot be understood other than as one element in a social structure, and that these societies have developed very different social structures, to a considerable extent based on differences in the social organisation of residence, as the products of different dominant ideologies. Arguing that 'Unfortunately, urban sociology has yet to come to grips with this diversity, or even to describe it adequately' (Kemeny, 1981: 62), I was critical of the continuing relative blindness of the new urban sociology to the problem of explaining diversity (pp. 62–3). This criticism remains valid today, as the overview of comparative housing research presented in Chapter 4 clearly shows.

Unfortunately, the theoretical underpinnings of this argument were weak and poorly explicated. In the first place, my criticism of the failure on the part of the new urban sociology to come to grips with the diversity of urban form was not related to issues of either unilinearity or convergence/divergence. Yet nowhere did I discuss social evolutionism or the convergence–divergence problematic, even though I was aware of the literature, and it had indirectly inspired the comparative approach I adopted.

Second, I left unexplicated the contrast that formed the central focus of my argument: between societies with well developed and poorly developed welfare states. The distinction which I used to understand this was described in terms of 'privatism' and 'collectivism'. This characterisation is very ambiguous and unclear. What about the situations where there is private provision but participation is compulsory, or there is public provision from which opting out is possible? There is clearly much to be done to make the concepts operational. In the last decade considerable work has been done by others on the issue of privatisation, which highlights the complexity of the issues involved (Lundqvist, 1989a, 1989b; Saunders, 1990).

Third, I did not attempt to theorise about any of the fundamental questions that underlay the thesis. Thus, for example, there was no discussion on the issue of the autonomy of the state, nor did I attempt to draw on existing theories of power, such as élite theory, class theory, or corporatism. Even more unsatisfactory was the absence of any reasoning around the concepts of ideology and myth which are so central to the argument, the latter even figuring in the title of the book.

Fourth, I failed to underline the difference between the approach I was using and that of traditional welfare state research. The impression was therefore given that my concern was limited to the relationship between housing and the welfare state, whereas I was attempting to go beyond a narrow welfare state approach and examine privatism and collectivism in the wider social structure.

Finally, I have been unwilling in my work to express the relationship observed between ideology and divergent social structures in terms of explicit theory. I contented myself with arguing that there exists a 'wider constellation of social values which are mutually reinforcing' (Kemeny, 1981: 64), without discussing how or why this comes about or what the causal factors are.

All of this is firmly in the traditions of empirical, policy-oriented housing research. In order to overcome some of these weaknesses, in the following section the arguments, are restated in more explicit form and some of the concepts clarified.

WELFARE STATES AND SOCIAL STRUCTURES

'Deep' and 'shallow' collectivism

The basic thesis underlying *The Myth of Home Ownership* is that there has been a process of divergence within the small group of industrialised societies between collective and private forms of social structure. This difference has little to do with the basic process of industrialisation, which all these societies have gone through. Nor is it simply a matter of certain societies having a well developed welfare state while others do not. Rather, collective and private forms of industrialism must be understood as different forms of social organisation at a more fundamental level.

My concern was therefore not only with the traditional welfare state areas such as housing, health and social insurance, but was also more general, involving many aspects of social structure, including the organisation of urban areas, transport, life styles, and gender roles. An important element of my thesis was that highly developed welfare states tended to possess also collectivised social structures, and that such collectivism complements and strengthens collectivist welfare arrangements. By the same token, poorly developed welfare states tend to be characterised by privatised social structures.

The distinction between collectivised state welfare and collectivised social structures is important and needs elaborating. As we have seen in the previous chapter, labour movement theory tends to focus on the welfare state. Its supporters are clear about the differences that exist between, for example, Sweden and Australia in terms of the extent of state involvement in the provision of social security. What they tend to neglect, however, are the differences between the wider societies of countries such as Sweden and Australia. The focus on welfare tends to overshadow more profound social structural differences, which are in some cases harder to quantify but which are cumulatively just as important, if not more so. Indeed, I would argue that it is not possible to understand why one society has a well developed welfare state and another does not without understanding the dynamic between the welfare state and social structure that underlies this difference.

My argument that patterns of housing tenure are related to the welfare state was intended to highlight this fact. It was not merely an argument about housing and welfare narrowly defined, it was also intended as a claim that housing, because of its central importance to the nature of social structure and its pivotal role in state welfare provision, provides the crucial link between the welfare state and social structure. I argued that it is just as important to look at the degree of collectivisation of the wider social structure as it is to look at the collectivisation of welfare provision. More specifically, not all social structures are equally conducive to the establishment of, and the continuing widespread support for, a highly developed welfare state. I therefore concluded that collectivised welfare systems are likely to be found in conjunction with social structures that exhibit relatively high degrees of collectivisation in their social organisation.

The idea that certain forms of social structure are more conducive to the development of a welfare state than others takes us beyond the dominant theoretical perspectives on the welfare state, such as labour movement theory and corporatism, and yields deeper sociological insights into the structure of industrial societies. It suggests that welfare states cannot simply be grafted on to any social structure with the same degree of success. Rather, the long-term viability of a welfare state varies in relation to the degree of collectivism of the social structure. This can be expressed in a simple two-by-two format, in which the degree of state welfare provision is matched against the degree of collectivisation of social structure (Figure 7.1).

Collectivised social structures with a high degree of state welfare provision produce the relatively powerfully entrenched welfare state model, as exemplified by Sweden. The polar opposite of this – a low degree of state welfare provision in a society with a highly privatised social structure – produces the liberal state model as exemplified by Australia.

The two 'unstable' combinations – welfare states constructed on relatively privatised social structures and liberal states in societies with

Figure 7.1 State welfare provision and social structure

Social structure	Degree of state welfare provision	
	High	Low
Collectivist	*Deep collectivism* Welfare state (Sweden)	*Shallow privatism* Unstable liberal state (Switzerland)
Privatist	*Shallow collectivism* Unstable welfare state (New Zealand)	*Deep privatism* Liberal state (Australia)

relatively collectivised social structures – represent 'shallow' forms of collectivism and privatism respectively. We may therefore distinguish between 'deep' collectivism and 'shallow' collectivism. The latter represent degrees of state provided welfare which do not match the degree of collectivism or privatism in the social structure. They can be described as 'unstable' in the sense that there is a mismatch between the extent of collectivisation in social structure and the extent of collectivism in state provision.

Expressing this in terms of hegemony and political dominance, as discussed in the previous chapter, shallow forms of state welfare organisation result from political dominance which has not been based on hegemony. That is, it is tantamount to adopting a 'statist' strategy without first using hegemony to create changes in civil society that reinforce and anchor hegemony in appropriate forms of social organisation.

Welfare states are therefore only the political manifestations of collectivist ideologies that result from the attainment of political dominance. For a welfare state to be enduring and viable in the long run this has to be buttressed by hegemonic collectivism that reaches down into social structure through a wide gamut of social institutions.

Welfare states characterised by shallow collectivism have not had their welfare systems buttressed with collectivised social relationships in the wider society. Or, if they had been so buttressed in the past, hegemony has subsequently weakened or lapsed to such an extent that they are no longer so buttressed. They are therefore characterised by collectivised welfare systems grafted on to privatised social structures. Such welfare states are likely to be unstable, with low degrees of congruence between the wider society and welfare forms, as a house that is built on sand. Welfare states characterised by deep collectivism are, by contrast, buttressed by forms of social organisation that complement and reinforce collective welfare provision. In shallow collectivism, therefore, the dominance of privatism in social structure creates social conditions which are not conducive to extensive state welfare provision. In such societies collectivism has taken a predominantly narrow political expression, and the social construction of a privatist hegemony could easily result in the collectivist political dominance being challenged. In shallow privatism the reverse holds.

Cases of shallow collectivism and privatism are particularly interesting. In a study of home ownership in Australia (Kemeny, 1983a), I argued that the relationship between high owner occupation rates and high levels of public welfare provision, while strong, was not without its exceptions. I discussed two societies which diverged from the dominant pattern: New Zealand, where the tenure pattern and social structure is similar to Australia but where the welfare state had been considerably better developed, and Switzerland, which possesses a large cost-rental sector and a small owner occupier sector but which also has a poorly developed welfare state (Kemeny, 1983a: 84–7).

I argued that the case of New Zealand suggested that the relationship between high rates of owner occupation and low rates of public welfare expenditure was more flexible than the general international correlation suggests, and that it was possible to 'stretch' the welfare state within certain limits in countries with high owner occupation rates. The case of Switzerland demonstrated that a large cost-rental sector with its associated collectivising effects on social structure does not have to be associated with a well developed welfare state, and that, while owner occupation with its skewed life-cycle housing costs placed limits on the extent to which the welfare state could be developed, a dominant cost-rental housing system was more flexible in that it facilitated the development of a welfare state but did not require it.[1] Since then, the dismantling of the New Zealand welfare state by both left and right governments (Davidson, 1989) would seem to support the thesis that, in highly privatist social structures, welfare states as expressions of shallow collectivism are unstable and liable to be undermined either by the absence of a dominant ideology grounded in deep collectivism, as would seem to have been the case in New Zealand, or by the social construction of a counter-hegemony, as may be happening in Britain.

There is an apparent contradiction in this analytical framework. It might seem odd that social structures which are highly collectivised should be compatible with a strong welfare state since it could be argued that it is precisely under those conditions that the need for a welfare state is redundant, or at least greatly reduced. The welfare state was developed largely as a result of working class pressure precisely because of the way in which industrialisation had disrupted traditional collective forms of welfare provision. Presumably, had working class grass-roots co-operation been strong enough, a syndicalist mutual aid or co-operative system would have made such demands unnecessary.

But this is precisely the point. The First International rejected the anarcho-syndicalist programme; henceforth, the aim of political strategy across almost the whole political spectrum from revolutionism to reformism, in most countries including Britain, became that of controlling the state. This took the form in most mass working class movements of a reformist strategic alliance with other groups, notably middle class reformers, in order to attain political dominance which would be used to create a strong central welfare state. The strategy in most countries was thus to bypass the arduous task of constructing hegemony based on deep collectivism and go for political dominance in one go. This is precisely the political lesson that Gramsci was trying to drive home. If the statist strategy had not been adopted, deep collectivism might have been constructed instead of concentrating on state welfare provision and the result might have been a decentralised welfare state along co-operative lines. But this is not the social reality we are faced with, and therefore the problem does not arise. Modern industrialism is based, for good or ill, on a considerable element of deep privatism *even in relatively highly collectivised social structures* (as we shall see in the Swedish case study in the following chapter), partly as a result of a century or more of the socialist – and it might be added communist – strategy of concentrating on state welfare provision rather than developing social structural mutual aid.

The reasons for the high degree of privatism in industrialised societies is therefore, ironically, precisely the strategy of using political dominance to build a welfare state rather than construct deep collectivism. But the very act of creating a welfare state had the effect of impoverishing deep social structural collectivism by removing key collectivist dimensions - such as social security and the provision of social housing - from the hands of ordinary people and centralising these tasks into state agencies. The result was that a privatist sphere was created into which ordinary people retreated, relieved of the task of creating collective grass-roots means of providing these services amongst themselves.

The distinction between collectivist and privatist social structures as presented in Figure 7.1, may therefore be refined by further distinguishing between moderately and highly collectivist or privatist social structures in relation to the role of the state, as in Figure 7.2.

Highly privatised social structures are those in which the dominant ideology is so strongly anti-collectivist that little or no co-operation is possible. The closest approximation to this is perhaps the 'amoral familism' described by Banfield (1958). Moderately privatised social structures approximate to the anglo-saxon model in which state welfare provision is

Figure 7.2 Collectivism and privatism and the role of the state

Degree of collectivism/privatism in social structure	State welfare role
Highly collectivised	Welfare co-ordination
Moderately privatised	Comprehensive welfare provision
Moderately privatised	Residual welfare provision
Highly privatised	None

minimal, while moderate collectivism approximates to the Scandinavian model. Highly collectivised social structures would be those in which co-operation is extensive and well developed, and in which the role of the state would be limited to co-ordinating the various co-operative activities and providing the ground-rules and laws to facilitate co-operation. Because highly privatised and highly collectivised social structures are rare, they can virtually be ignored for the purposes of developing a divergence thesis.

The degree of collectivism/privatism is therefore a complex phenomenon, the form of which evolves from the interaction between the nature of social structure and the organisation of the state. In order to develop the argument, I start by clarifying what I mean by (moderately) collective and private forms of industrialism. I then explain why it is that housing – or, more accurately, 'residence' – constitutes such a pivotal dimension which plays a major part in determining the specific mix of collectivism and privatism in a society, in terms of both state provision of welfare and the wider society.

Collectivism and privatism: a framework

We noted in Chapter 5 that welfare can be provided in many different forms; notably the state, voluntary organisations, co-operatives, for-profit organisations, employers, mutual aid between households, and within households. These represent different forms of social organisation for the provision of welfare. Some involve more elaborate social structural arrangements than others, while there are also differences between them in terms of the fundamental principles under which they operate. But how do these huge variations emerge?

The concept of social differentiation as developed in the classical sociology literature is useful here. Societies become more complex and new forms of social organisation emerge as specialisation through the division of labour becomes more intense. This dynamic can easily be seen at work in terms of the provision of welfare.

The simplest form of welfare provision is that within households, which can be extended and buttressed by drawing on kinship, neighbourhood and friendship networks in order to set up inter-household co-operation, or mutual aid. Taking childcare as an example, this can take the form of childminding by grandparents or neighbours, or, with greater complexity of organisation, childminding circles. These in turn can develop more formal co-operative childcare organisations, or they may form the basis of for-profit organisations, or even charitable organisations, perhaps developing into voluntary organisations. These formalised and complex organisational forms of welfare provision can also be socialised and provided directly by a stage agency.

The paths between these different forms are potentially numerous and complex. They include welfare provided by for-profit organisations for their own employees only, or by charitable organisations operating in selected localities or among certain categories of the population. Finally, hybrid types of organisation somewhere between a for-profit and co-operative organisation (for example, a church organisation) may provide welfare in various forms. The provision of welfare in social structure must therefore be understood as a veritable kaleidoscope, a constantly shifting pattern of social institutions, highly complex and inherently unstable in the sense that it is open to innovative change.

It should be noted that household provision constitutes the basis from which privatist or collectivist forms of social organisation emerge. In this sense, then, household provision is fundamental and contains within it the potential for social differentiation in a wide range of different forms. It therefore constitutes a potential source of either collectivisation or privatisation. Changes between state, voluntary, household, and market in any direction affect the overall balance between collectivism and privatism (Figure 7.3).

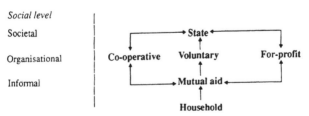

Figure 7.3   The social differentiation of welfare provision

This dynamic of the balance between private and collective is a fundamental dimension of social structure. It is normally not studied directly, even though the individual–society tension which underlies it has been a central problematic in sociology (for example in the *gemeinschaft* and *gesellschaft* distinction and primary and secondary realationships). However, it provides a revealing angle on the nature of social order. The general dynamic of how this balance is attained – and how it may be changed – is often understood in the sociology literature as a function of the logic of development, an explanation which has its origins in social evolutionism. I want to suggest here that the process of differentiation is not so simple and unproblematic. Nor is it in some sense beyond human agency. Rather, it can be understood as the product of the social construction of hegemony – of social action, as individuals struggle in their interpersonal interaction to create new forms of social organisation. It is in this process that ideology can provide a powerful means of fostering support for one form of social organisation over others and thereby to channel social differentiation along certain lines; for example, towards more for-profit companies as against co-operative enterprises, or towards collective as against privatist forms of residence. Struggles over degrees of collectivism in social structure and in welfare therefore play a vital part in determining the balance between privatism and collectivism.

Certainly, important ingredients in producing new patterns of social interaction and organisation are such factors as new technologies, war, and economic cycles. But fundamental to both the direction of change and its depth is consciousness and the way in which ideology channels energies into the social construction of one or another dominant type of social organisation. Seen in this way, social structures are no longer unyielding set-in-concrete phenomena but are, in fact, highly malleable and are sensitive to changes at many different levels. Indeed, they are, as indicated above, in a constant, kaleidoscopic, process of flux, particularly at the household level, and the potential for the emergence of different forms of collectivist or privatist social organisation is great. At the same time, fundamental reorganisations of social structure are longer term processes that involve more far reaching changes in the larger and more complex organisations of society. It is in this sense that divergence must be understood as a process taking place over several decades.

Fundamental to this process are the ways in which households relate to one another to construct either privatist or collectivist forms of social organisation. Deep collectivisation would involve intense co-operation at the household level, while less intense co-operation results in less deep forms of collectivism. Thus, for example, Banfield showed how a strong individualist ethos could, under certain social conditions, act as a hindrance to industrialisation by preventing the co-operative effort between households which would enable surpluses to be accumulated as the basis for investment in development.

Taking the example of residence, collectivism is well illustrated in the early stages of the urbanisation of the USA through the work of Hayden (1981). Hayden sets out to show how feminists and a wide range of left and utopian thinkers and community builders attempted to develop forms of housing that provided high levels of collective facilities. The reasons for this impetus were diverse, as was the degree of collectivist residential provision envisaged. Among feminists, for example, the intention was to free women from much household labour by sharing it. Among other groups the concern was to develop communes. Among yet others, ideas of rationalising domestic labour predominated. But the strands of thought all ultimately derived from a deep-rooted dismay at the kind of isolated nuclear familism that was emerging out of the suburbanisation process, combined in some cases with a continuing tradition of anarcho-syndicalism. The schemes never took root. The movements that sponsored them remained splintered and ineffective, never becoming part of a major alliance and never therefore taking part in the social construction of an ideology that could mount a challenge to become hegemonic. Instead, the suburban ideal of separate households became incorporated into the dominant ideology and ultimately became all but universal.

We will see later in this chapter and Chapter 8 how Swedish residence exhibits a greater degree of collectivism than is common in many industrialised societies, and how this came about as a result of the construction of social democratic hegemony and political dominance. The point to note here is that, irrespective of tenure and, to some extent at least, of dwelling type, residence can vary considerably in its degree of collectivism. Of the myriad of possible ideas for organising residence, one tends to emerge as dominant. The crucial ingredient for success is a movement which is able to establish a hegemonic ideology in which residential organisation figures as a major element.

Given this approach, the spectrum of the social division of welfare can be analysed in different ways. Esping-Andersen (1990) chooses to make a threefold distinction between state, voluntary/employer, and market. Here, I propose to differentiate a collective/private spectrum in which a society is more collectivised the more state and voluntary provision there is and more privatised the more for-profit provision there is. In choosing this approach, if I understand Esping-Andersen's framework correctly, I effectively subsume his corporatist form of welfare capitalism under the other two, though the precise fit remains unclear. This is not to be taken to mean that I reject his 'trifurcating divergence' model. Rather, just as the difference between unilinealism and divergence is a matter of levels of analysis, so here I choose a somewhat higher level of generalisation: one that lies between unilineal and multilineal levels. This can be illustrated in terms of housing, which is a particularly important link between the welfare state and wider society. A society with highly collectivised forms of housing is more likely to have also a well developed welfare state precisely because the embeddedness of housing in social structure means that its collectivised character will impart a considerable degree of collectivism to social structure. By the same token, highly privatised housing systems will counteract and undermine collective welfare by its privatising effects on social structure.

In Kemeny (1981), home ownership was used as an *index* of privatism in housing, with public renting being a measure of collectivist housing, in much the same sense as the use by Durkheim of suicide rates as an index of social solidarity. This had one unfortunate consequence in that it focused upon tenure issues at the expense of the privatism/collectivism dimension of which the tenure categories were intended to be crude measures. Breaking down the dimension into component parts that cut across tenure categories is thus an important conceptual refinement that needs to be made.

It is therefore necessary to move away from the simplistic identification of owner occupation with privatism in housing, and to develop a measure of privatism in housing that includes other forms of housing than owner occupation, using the above continuum of privatism/collectivism in housing forms. Highly collectivised housing forms would then be those in which the production, financing, distributing, and maintenance of housing were organised on a collective basis by the state or by some voluntary form of provision or by tied (employer provided) housing or by a characteristic mix of these.

But, over and above this, some forms of housing might by these criteria be relatively privatist, *irrespective of the type of provision*. Thus, for example, public renting where housing is financed out of taxation, is built using direct labour organisations, and where strict non-profit principles using rent-pooling is applied, is more collectivised than public renting where housing is financed on the market, is built by private companies, and where rents are high enough to generate surpluses which are then extracted, or where rents are set at individual historic cost levels and tenants are allowed to sell access to such housing. It is also possible to conceive of a voluntary housing sector, made up of non-profit organisations, which fulfils many of the criteria of a social market but with little or no state provision.

It can therefore be seen that issues of collectivism and privatism are highly complex, and that simple dichotomies in terms of, for example, owner occupation as against cost-renting, such as I have used in earlier work, are gross oversimplifications that can sometimes obscure more than they reveal. This is an area that has been almost entirely neglected in housing research, and where comparative studies of, say, public renting might yield surprisingly large variations in the degree of collectivist principles that operate in the sectors. It is therefore important to be aware not only that public rental housing – and indeed other forms of tenure – in different societies vary in the degree to which they are collectivised but also that the combinations of different forms of tenure in different countries, each manifesting different degrees of collectivism, result in the housing systems of some societies being in some overall sense more collectivised than others.

Finally, and perhaps most important of all, the differences in the social division of welfare that exist between different social groups and different localities but within formally identical tenures may be more significant than the differences between tenures and housing forms in their effects upon social inequality. For example, in terms of social differences, house buying among some ethnic groups may involve alternative forms of finance comprising mutual aid based on kinship borrowing. Similarly, the degree of collectivisation may vary significantly between different local authorities.

The resulting judgements that have to be made in terms of the degree of collectivism/privatism in particular housing systems are therefore potentially complex. This complexity is greatly magnified when the criteria are applied to other sectors of society in addition to housing, in order to arrive at an overall estimate of the extent to which a society is collectivised. Such complexity may well be unquantifiable. It ultimately rests on an intuitive and many-dimensioned understanding of the collective-private dynamic, and the ways in which the balance between them is struck in different societies. In the following discussion and in the absence of empirical data, I do not attempt to apply the above framework in any systematic manner but only to use it as a rough guideline.

COLLECTIVE AND PRIVATE MODES OF INDUSTRIALISM

In The Myth of Home Ownership I argued that, on the basis of crude international comparative welfare and tenure statistics, it was possible to distinguish between dramatically different levels of state and civil society collectivism. The USA and Australia represent examples of highly privat-ised social structures, while Sweden, The Netherlands, West Germany, and the former state socialist societies of Eastern Europe represent examples of societies with varyingly high degrees of collectivism. In between these lie a large number of societies which combine collectivist and privatist elements in various proportions and which fall somewhere between these two purer forms. Three case studies were chosen which exemplified the two polar types of privatism (Australia) and collectivism (Sweden), with a third intermediate case (Britain).

I argued that Sweden is an example of a society in which a high degree of collectivism has been established, in terms of a well developed welfare state, highly developed collective transport based on high density urban form derived from a dominance of rental and co-operative flats, and public childcare facilitating predominantly wage-labour female roles. By contrast, in Australia there has evolved a high degree of privatism, as reflected in a poorly developed welfare state, low residential densities deriving from privatised dwelling type and housing tenure, private transport, and predominantly 'domestic' female roles. Britain represents an intermediate form of social organisation, with a welfare state more developed than Australia but less so than Sweden, more social housing than Australia but less than Sweden, predominantly semi-detached and terraced housing rather than either freestanding villas as in Australia or apartments as in Sweden, female roles neither as domestic as in Australia nor as wage-labour based as in Sweden, and so on.

The differences between these three societies are striking. Moreover, a central feature of the differences is the role of housing, particularly in terms of form of two major dimensions of housing – tenure and dwelling type. The data presented in *The Myth of Home Ownership* comprise two main categories: data on the welfare state and its relationship to tenure and data on the relationship between dwelling type (house versus flat) and social structure. We may consider these in turn.

The interaction between welfare and housing tenure

Data on the welfare state and tenure comprise very general statistics on per capita indices of expenditure on various categories of government spending, and levels of taxation, for a number of countries with both high and low home-ownership rates, supplemented by similar but more detailed data on Australia, Britain, and Sweden. The data show quite clearly that the former had low rates of spending but high owner occupation rates, while for the latter the opposite was the case, and Britain lay somewhere in between both in terms of public expenditure and owner occupancy rates (Kemeny, 1981: 57–8, Tables 4.3, 4.4; see also Kemeny, 1980).

On the basis of these data, I speculated over the role of housing in relationship to social security and welfare. I argued that housing comprises such a large item in the household budget that tenure differences will have profound implications for resistance to or acceptance of public collective intervention. I hypothesised that the skewed lifetime housing costs of owner occupation, whereby the bulk of the costs are concentrated in the early years of the family cycle rather than spread evenly as in renting, functioned as a structural deterrent to both high taxes and high levels of social security. The highly individualised household burden of housing in owner occupation will thereby act as a deterrent to the implementation of equalising mechanisms of the welfare state because housing costs absorb such a high proportion of the household budget that it will generate resistance to collective arrangements in other areas.

This is best illustrated with two examples: health insurance and retirement pensions. Health insurance represents a means of redistributing health care from the young, childfree and healthy to the old, to households with children and to the sick. High housing costs tend to occur at the early stage of the family cycle when health tends to be good and before the childbearing stage. Because of the pressure on household budgets first to save for a deposit and then to finance high mortgage repayments, there will be low resistance to the introduction of voluntary contribution health insurance among younger households. By the same token, there will be less pressure on politicians to introduce high retirement pensions because the living costs of the elderly in a society where owner occupation is widespread, or even the norm, can be subsidised by low housing costs resulting from both zero mortgage debt and the running down and neglect of maintenance (see Kemeny, 1981: 55–62; 1977).

A clarification of what this implies is necessary at this point. The argument is not that households will necesarily hold such attitudes: for example, that the welfare state is undesirable or less necessary, as, say, Saunders (1990) assumes. Saunders argues against my thesis by citing attitudinal data from surveys carried out in one country only (Britain) to show that owner occupiers do not hold anti-welfare state attitudes. However, the effect is more indirect, in that it facilitates political decisions to be made which involve reducing state welfare commitments without encountering strong public opposition. There is no direct and simple correlation between survey attitudes and policy outcomes. Rather, the latter are mediated by hegemony and its social construction. This is why, for example, politicians in Australia have been able to argue successfully against large pension increases. They have been able to resist political demands for higher pensions, made on the grounds that tenants live below the poverty line, by arguing that the majority of Australians are owner occupiers, and that households who chose not to buy must accept moral responsibility for their situation and put up with the resultant high housing costs in old age, rather than be baled out by the state (for evidence see Kemeny, 1977: 51).

It is important to note the implication of this. There is a subtle interplay between existing degrees of collectivism or privatism in social structures and the attempt to construct political dominance. The argument that high pensions are not necessary because most people are owner occupiers builds upon the prior existence of deep privatism. Deep privatism in social structure is thereby reinforced by political measures. This is not to say that a challenging hegemony cannot be assembled, for example by a labour movement building a strategic alliance to do so, or that, if this were done, an alternative political dominance could not be attained which could support more collective forms of housing. Essentially, then, the relationship between tenure and welfare is mediated by a complex of political and social factors and cannot be explained in simplistic and inter-tenural attitudinal terms within one country.

Since my earlier work, further data of better quality have strengthened the evidence of a close relationship between housing tenure and the welfare state (Schmidt, 1989a) and there is now, I believe, a much stronger prima-facie case for the thesis. The data are in the form of correlations, the most convincing being that between public social security expenditure, as a percentage of total government expenditure, and percentage of owner occupation for 17 countries, yielding a correlation r of (-0.90). Another correlation – between public social security expenditure expressed as a percentage of gross national product and percentage of owner-occupation for 17 countries – yields a correlation r of (-0.83). These are high correlations by any standards, and provide a very promising starting point for more detailed empirical investigation of the relationship between residence and the welfare state.

It is clear that the relationship between housing and the welfare state needs much more comparative work in order fully to understand its dynamics. However, the data of Schmidt (1989a) and of Esping-Andersen (1990) only really address the issue of state welfare. They both show only that different industrial societies have different levels of state welfare provision while Schmidt in addition shows that this is correlated to housing tenure. But the thesis posited in Kemeny (1981) was not only about the welfare state and housing tenure, although this did comprise an important part of the picture. The argument was rather that the *total social organisation* of these societies differs fundamentally from one another, and that housing plays a key part in constituting the differences. The second category of data, on broader forms of social organisation, was therefore at least as important – indeed in some respects more so – than the welfare state data.

The interaction between spatial form and dwelling type

These differences centred on differences in dwelling type between the three societies, particularly in urban areas. The differences are quite dramatic.

Seventy per cent of dwellings in Greater Stockholm comprise flats, compared to only 22 per cent in Greater Sydney (Kemeny, 1981: 44, Table 4.1). The overwhelming majority of dwellings in Australian cities are houses – indeed, detached houses, mostly bungalows in quarter acre plots – while the overwhelming majority of dwellings in Swedish cities are flats. Furthermore, Britain, as in terms of state welfare, falls squarely between these two extremes, with 45 per cent of Greater London comprising flats. In addition, the overwhelming majority of London houses are terraced or semi-detached, not detached as in Sydney or Stockholm. The contrast in city housing is clear.

The difference this makes to the socio-spatial organisation of the cities of these countries is marked, and has been discussed in Kemeny (1981: ch. 4). It profoundly affects, for example, the balance between public and private space (e.g. parks and gardens), between public and private modes of transport and between domestic and wage-labour female roles. This suggests that the single difference between societies of the predominance of one dwelling type over another can have such profound consequences for the social organisation of everyday life that this fact alone could possibly constitute the basis for understanding divergence between industrial societies. At the very least, the social organisation of residence must be a major factor.

As admitted in Kemeny (1981), the empirical underpinnings of this spatial dimension of divergence are weak. Data on differences between cities with different spatial form are rare. One of the few comparative empirical studies I was able to find was that by Szalai (1972), an international survey of the uses of time. From this I was able to isolate two cities with very different spatial structures: Osnabrück in Germany, with a population of 138,000 and where 68.2 per cent of dwellings were flats, and Jackson, USA, with a population of only 72,000 and where 82.9 per cent of dwellings were one- or two- family houses. In Osnabrück 96.2 per cent of dwellings were within 5 kilometres of the city centre, compared to only 48 per cent in Jackson, with little more than half of Osnabrück's population. Further, whereas only 16 per cent of Osnabrückers drove to work, fully 63 per cent of Jacksonians did (Kemeny, 1981: 50).

Sociological studies of comparative urban form are also rare. The most relevant is Popenoe's (1977) comparison between a US and a Swedish suburb: Levittown outside Philadelphia and Vällingby outside Stockholm. Popenoe contrasts the low density and car-dominated US suburb with the high density Swedish suburb, and this detailed and careful study brings out clearly the impact of different urban form on life styles, particularly in respect to the role of women.

Data such as these provide circumstantial evidence for a divergence thesis, and it is likely that considerable amounts of such data do in fact exist in different contexts, for example in town planning archives and other surveys which touch incidentally on the question of urban form. However, the traditional concerns of urban sociologists have been in delineating and understanding the social consequences of urbanism in general, rather than in developing a divergence thesis of urban form. This has manifested itself in a comparison between urban and non-urban and in a quest for the consequences of urbanism as a general phenomenon for modern society, in a tradition that includes just about every major social theorist since Comte (for overviews see Saunders, 1986; Smith, 1980).

However, since then, there has been a very substantial growth in interest in space and its social structural correlates (see Duncan and Savage, 1989 for a review), though as yet most energy has been spent on analysing locality differences, and these are usually locality differences within rather than between nation states. One of the few theoretical studies of the effect of space upon social structure is that by Hillier and Hanson (1984). They argue that the physical arrangement of buildings and public/private spaces reflect relationships of power and control and so must be understood as expressions of ideology (Hillier and Hanson, 1984: 260–1). But this, too, deals with small-scale phenomena. What has not yet been addressed in any systematic way is the effect of very different kinds of urban form upon social structure, particularly cross-nationally. Yet, surely, differences of the magnitude of those between the spatial and social organisation of two rich and highly urbanised industrial societies such as Sweden and Australia should constitute the basis for a research agenda of prime importance?

CONCLUSIONS

My argument is therefore that divergence must be understood in broad social structural terms, and not merely in terms of the welfare state. The welfare state represents one way of organising welfare, using the state as a means of attaining goals of social equality. But the development of the welfare state cannot be understood in isolation from structures in the rest of society, and particularly in relation to broader tendencies towards collectivism or privatism in society. These tendencies will be powerfully influenced by the organisation of residence. This will manifest in a number of ways, but crucially in two important respects: the social forms which emerge around the ownership of, and household financing of, housing (usually expressed in terms of tenure), and the spatial consequences of the dominance of one or more dwelling type, particularly in urban areas.

The patterns of privatist and collectivist social structures indicated by the above are not necessarily fixed and immutable. Over longer historical periods it is possible for the emphasis within a society to shift quite markedly. For most of the postwar period, and indeed in some respects since even earlier, the very considerable expansion in the welfare state has been an important factor in encouraging a growing tendency in many of these societies towards increasing collectivism. In the last decade or so, however, a number of societies (notably Britain and most dramatically former state socialist countries of Eastern Europe) have shown a tendency to move towards increasing privatism, particularly in terms of housing but also in a more general withdrawal of public welfare provision. The shifting constellations of privatism and collectivism among industrial societies may be one reason why convergence becomes a research issue at certain times rather than others.

The purpose of this discussion has been to begin to specify the elements of a divergence thesis in comparative housing by developing a more theorised version of *The Myth of Home Ownership*. The whole requires better anchoring in the theoretical literature, with the intention of redefining the research of housing issues away from narrow 'housing studies' towards wider social structural questions. There are, of course, a number of different elements to this task, and these have been raised in this chapter.

The first task is to move the somewhat restricted debates in comparative housing studies from their present unsatisfactory state – undertheorised and limited to bringing out variations in structures that can only be explained in terms of random historical contingencies or stages of development - towards asking more systematic questions about differences between societies.

The second task is to sharpen the basic thesis developed in *The Myth of Home Ownership*, in which housing is placed in a wider social context and a divergence thesis is posited which is based on privatist as against collectivist forms of industrial society. Important implications of this are that narrow studies of welfare states are inadequate and that highly collectivised welfare provision cannot be understood separately from the degree of collectivisation of social structure. Central to this is housing which provides a key link between social structure and the welfare state.

Beyond this, the development of a divergence thesis can contribute to breaking down the powerful image that societies are the helpless victims of forces of evolution. It can begin to indicate in major ways the manner in which directions of social change are more open than unilinear and developmental theories suggest. The interplay of social, political and economic forces, and the roles of the different interests in these, emphasises the inherent instability and unpredictability of social structure, the fact that societies can be changed by the intervention of human agency, and the major role that such agency can and does have in moulding social structures.

8 The political construction of collective residence: the case of Sweden

INTRODUCTION

In this chapter I want to draw together some of the threads of the discussion in the preceding two chapters and present some evidence in support of a divergence thesis. This evidence is limited to one aspect of such a thesis, namely the social construction of public policy towards the organisation of residence. By residence, I mean households-in-dwellings situated in their local urban, rural, or regional contexts. This must be seen as just one part of a broader process of social construction of the wider society in which social order is negotiated. The framework used can be broken down into the following elements:

1 The constitution of society develops through broad social movements and their conflicting ideologies as they struggle to establish, counter, or defend hegemony.
2 The process whereby this is achieved is in the social construction of negotiated orders in which modes of discourse and frames of debate are established. These take place in a wide variety of contexts and hegemony is achieved when one group manages to determine the mode of discourse and to use it to establish a dominant ideology in a number of key areas.
3 Hegemony is converted into political dominance by means of developing a strong position in relation to the state and state power, normally through holding governmental office for long periods.
4 Housing – or more broadly residence – comprises a key arena in which a hegemony is established, owing to its impact on social structure through its salience and its embeddedness in broader social issues.
5 Housing policy, by the same token, comprises a key area in which hegemony can be converted into political dominance which in turn can frame and influence basic characteristics of housing, such as tenure and

its impact on household consumption and dwelling type and its impact on urban form.

6 Hegemony is effectively converted into political dominance in issues of residence by entrenching myths, which are consonant with the dominant ideology, into the negotiated orders of public policy within the relevant political arenas. Such 'agenda-setting' comes to frame the major political debates and to channel change in directions consonant with the dominant ideology.

7 Divergence between countries takes place as a result of the establishment of contrasting hegemonies based on very different dominant ideologies. These develop over long periods of historical time: decades and generations, rather than just several years. Residence comprises a key dimension of this.

In this discussion, I want to examine the establishment of political dominance and the role that myths have played in the formulation of, and support for, residence policy in Sweden. As argued in my earlier work, Sweden has a radically different approach to housing from Britain and Australia, which derives from more general attitudes and values and results in different models of modern industrialism. Its primary characteristic is the higher degree of collectivism of Swedish social structure compared to most industrialised societies. This chapter, therefore, examines some of the ways in which public policy towards residence has contributed to the development of 'deep collectivism' (see Chapter 7).

This pattern of divergence must be understood in terms of the elaboration and establishment of contrasting ideological frameworks in which different myths inform public policy and provide the emotional energy behind the nature of state involvement. The case study therefore shows the importance of contrasting myths behind the establishment and defence of particular kinds of public policy within broader and contrasting ideological contexts, which themselves must be understood as constituting elements in the social construction of hegemony.

THE SOCIAL CONSTRUCTION OF HEGEMONY

There is a large literature on the Swedish welfare state, as interest has been considerable in the strong social democratic movement and its impact on the organisation of welfare (Castles, 1978; Davidson, 1989; Esping-Andersen, 1985; Heclo, 1974; Heclo and Madsen, 1987; Korpi, 1978, 1983; Wilson, 1979). What is striking about this literature is that, despite the generally high awareness of the importance of ideology in the creation of the Swedish welfare state, awareness of the potential relevance of the concept of hegemony is almost totally absent.

A good example is that of Castles (1978) who is particularly sensitised to a class/ideology perspective in the Swedish context. He argues that where the capitalist class is unable to dominate the political process, as in The Netherlands and Scandinavia, welfare and egalitarianism may come to occupy an important place in the dominant image of society (Castles, 1978: 97). He argues that 'the creation of a social democratic image of society may in itself be a socialist achievement of some magnitude' (Castles, 1978: 97). Castles later argued that the reasons for the social democratic dominance in Sweden is to a large extent dependent on the fact that right wing politics are fragmented, and that the relative strength of social democracy is in part a reflection of the divisions of its opponents (Castles, 1982).

Like Castles, both Esping-Andersen (1985) and Korpi (1978, 1983) place the locus of their explanation of the strength of social democracy on the level of politics. But Esping-Andersen (1985), in particular, is aware of the need for the social democratic movement to construct and sustain a viable alliance between traditional working class and white collar workers. He argues that such a policy has underlain social democratic strategy from the start, but that under modern circumstances there are clear signs that this policy has been under strain and is in danger of collapse. Esping-Andersen terms this weakening of traditional alliances 'decomposition'.

All of these authors would have enriched their analysis by an understanding of the process of what I would term the social construction of hegemony and its translation into political dominance. Castles is probably most aware of the role of ideology in the creation of hegemony, while Esping-Andersen tends to focus most on the creation of political dominance through alliance building. A combination of the different emphases of these authors, held together by an overall theoretical perspective on hegemony and complemented by a negotiated order or other constructivist perspective, would constitute a useful starting point for a clearer understanding of the emergence of modern social democratic dominance in Sweden and its manifestation in terms of housing policy.

The outstanding – almost unique – characteristic of the urbanisation of Sweden was that it took place under a powerfully entrenched labour movement, represented by a Social Democratic party, which successfully assembled an alliance of disparate groups, the so-called 'People's (or Popular) Movements' (*folkrörelser*). This included the temperance movement, various housing movements such as those of tenants and of tenant-owners (which included significant bourgeois elements and had similarities to the British building society movement), religious movements, the electoral reform movement, and associations of both rural and urban workers.

The Social Democratic party thereby became the political expression of a wide range of interests extending beyond narrow working class interests, even if its principal source of support remained the labour movement.

This successful forging of a broad front alliance has had far-reaching implications for the ideological interpretation of the urbanisation of Sweden. It meant that it was possible to depict with considerable conviction the recent history of Sweden in terms of the success of social democracy in accomplishing the restructuring of the process of industrialisation. Social democracy was in a position to claim the credit for the modernisation of Sweden; the collectivistic principles on which the movement was in theory based could be held up as fundamental ingredients in a success story – irrespective of how much such principles were applied in practice. In this way the popular movements became part of the political rhetoric of modernisation in a way that did not happen in other countries, even where similar movements played a major part but failed to develop hegemony.

But the forging of a broad alliance was only part of the overall strategy. Equally important – possibly more so in the long run – was the creation of broad national unity including the representatives of capital. This is epitomised in the *Saltsjöbad* agreement in 1938 (named after the location of the talks) at which labour and capital organisations agreed to the ground rules for wage negotiations.

The result of this was also that social democratic ideology became truly national, incorporating broad themes to include interests of various component movements (such as 'democracy' and the reform movement), and also incorporating right wing interests. This cross-class co-operation has remained a major characteristic of Swedish political life. It created a corporatist system under social democratic hegemony, which I have termed 'hegemonic corporatism' (see Chapter 6) as it remained under social democratic leadership in which all parties at least adopted the same general mode of discourse.

The uniqueness of the Swedish social democratic strategy was, therefore, to include as many other social interests as possible, *including those of capital*, and to construct a truly national movement – unlike the Labour Party in Britain, which remained narrowly class-based and failed to assemble a broad alliance across class lines or to develop a dominant ideology. The attainment of hegemony was a central feature of this strategy. This therefore distinguishes it sharply from what might be termed 'labourism' with its fear of populist movements (such as nuclear disarmament and the environment), its narrow class basis, and its combative and oppositional attitude to capital and other bourgeois interests.[1]

The social construction of social democratic hegemony in which a language and perspective, including research and technological applications, came to be defined within certain limits and according to certain terms does not simply take place through political speech-making and official documents. Rather, these are the end products of a long process of interpersonal interaction taking place in numerous different social settings. In the case of Sweden, its starting point lies in a basic everyday social construction of reality, achieved through local networks, often in the form of union activities, but more often through an extensive and far reaching programme of evening classes or 'study circles' (*studiecirklar*).

Tilton (1988) identifies five basic ideological themes in Social Democratic policy. The first he termed 'integrative democracy', referring to the democratic road to socialism and, more important, the creation of a broad social front in alliance with other elements which transcends party and class boundaries. The second is the idea of 'the People's Home', about which more later. The last three are the belief that a basic precondition for 'efficiency' is equality, the belief in the social control of the market economy, and the belief that a 'strong society' and the security that it engenders increases, rather than decreases, individual choice.

Some of these elements run directly counter to dominant liberal/ conservative ideology, and constitute the basis for what might be termed a collectivist mode of discourse. Others are based on traditional, and often deeply conservative, rhetoric but have been given an opposite meaning. A key element in the collectivist rhetoric of the dominant mode of discourse was – and to a large extent still is – the central importance of the welfare state and the belief in planning as the solution to social problems. A key motif of this mode of discourse was the idea of 'the People's Home' (*folkhemmet*). This was described in a famous speech in 1928 by one of the founders of social democracy, Per Albin Hansson, later to be prime minister, as follows:

> The basis of the home is togetherness and mutual affection. The good home does not allow the existence of any privileged or deprived members, or any scapegoats or stepchildren. In the good home no-one looks down on anyone else, nor tries to gain at anyone else's expense, nor do the strong bully or rob the weak. In the good home equality, consideration, co-operation and helpfulness prevail. *Applied to the larger people's and citizens' home* this means the breaking down of all social and economic barriers which at present divide citizens into the privileged and the deprived, the dominant and the dependent, the rich and the poor, the propertied and the destitute, the exploiters and the exploited.
> (Cited in Tilton, 1988: 371; my translation, my emphasis)

This formulation of the welfare state, which came to be subsequently dubbed 'the People's Home', comprises two essential ingredients. The first is its broad appeal in terms of all-inclusiveness. The welfare state came to be defined as the nation-at-home, closely identifying nationhood with the welfare state. The rejection of a narrow formulation of the welfare state in class terms must be understood as an important advance from the politics of class to the establishment of hegemony based on the assembling of a broad alliance of interests. The second is the manner in which the People's Home is a clear example of a myth in the transposing of the family idyll and the ideal of the home – as the nexus of household and dwelling – on to how a moral and proper society should function: as one big happy family, ensconced within the four walls of its home. Hansson paints a picture of the ideal home and then transposes it on to the whole society.

It is particulary interesting that the concept of the People's Home has deep conservative and traditional roots going back to the almost unique pattern of the early industrialisation of Sweden in the numerous small one-factory villages run by the owner on paternalistic lines. What the social democrats did was to redefine the meaning of the People's Home by emphasising those stereotypes concerning the home as an idyll that matched most closely their own class ideology (such as equality and co-operation) and playing down other characteristics that would suit a more conservative ideology (such as paternalism and sexism, hierarchy and ageism, obedience and discipline – 'sparing the rod and spoiling the child' – and authoritarianism) that are just as valid as home stereotypes.

The idea of 'the People's Home' has been very powerful in the political rhetoric of policy-making (Carlestam, 1986: ch. 6; Hirdman, 1989). But it must be understood in the broader context of the social construction of a dominant mode of discourse. Central to this process was the emergence of a politically engaged intelligentsia not merely to formulate and express the ideology and develop a mode of discourse but to create the climate in which social planning, already a well developed science in Swedish politics, could become the basis for the construction of a welfare state. Ollsson (1990) has traced in some detail the development of what he terms 'a social policy discourse' in Sweden, in a wide-ranging discussion of the principal ideas of the leading thinkers around this issue, including Hjälmar Branting, Gustav Möller and Gösta Rehn. Perhaps the best known internationally of this radical intelligentsia were Gunnar and Alva Myrdal. Early on they were in the forefront of the argument for basing the People's Home on planning and rational social investigation.

This technocratic approach to welfare, with its social surveys, assiduous collection of data, and measurements and standards set down to the finest detail is particularly evident in housing policy. The periodic Living Customs Surveys (*levnadsvanundersökningar*: use of and attitudes to the home), and the space and standard construction norms, with criteria specified in terms of size down to the smallest kitchen cupboard, epitomise the detailed planning mentality that lay behind implementing the myth of the People's Home. This was manifested most dramatically in a whole series of public investigations into a wide range of topics from the 1930s to the present.

The careful planning of the People's Home marks a fundamental difference in perspective towards the welfare state than, say, in Britain. It generated a whole language of policy-making: catch-words that express the spirit of the enterprise. One such is the expression 'society-building' (*samhällsbyggandet*), the metaphor of 'building' fitting in well with that of a 'home' and emphasising the possibility of planning a society.[2] Society is not something that simply 'is', it is constructed, assembled, and ordered. Another is that of *planhushållning* (untranslatable except literally as 'plan-householding', being a combination of economic efficiency with public planning of 'the national household' based on principles of social equality). Yet another is *socialpolitik* the idea of 'social' politics (the word *politik* in Swedish means both 'policy' and 'politics' and therefore has a broader and more all-embracing meaning than policy in the narrow sense). By the same token, 'social policy' means more than policy in terms of social security. Rather, it reflects the whole spirit of the People's Home in every area, including, of course, housing where such a policy is termed a 'social housing policy' (*social bostadspolitik*). Adding the word 'social' to policy does not add any meaning whatsoever beyond the emotive connotations of the welfare state. It is a significant index of a dominant mode of discourse that even the most right wing party, the Moderates, talk about a 'social' housing policy in their party political manifesto. There are also powerfully emotive negative expressions which have no equivalence in English, such as 'category housing' (*kategorihus* or *kategoribostäder*) which loosely means segregated housing or housing built for certain social groups and which therefore violates the equality principle.

The construction of a dominant mode of discourse extends well beyond images of the welfare state. Swedish political language differs significantly from that of English-speaking countries in the continued existence of a terminology long since abandoned by socialists in most other countries. Terms such as 'bourgeois' and 'solidarity' which are archaic in the English language even on the political left still carry powerful emotive force in Swedish political discourse. There has even been a government commission of investigation report entitled 'A solidaristic housing policy' (*en solidarisk bostadspolitik*: SOU, 1974).

The dominance of this mode of discourse is best seen in terms of the extent to which it is adopted by those who oppose the ideology which the mode of discourse frames. Tilton underlines the extent to which this has happened in Sweden:

> However, ideology is most effective when it affects the opposition, first by dividing them and thereby facilitating the implementation of reformist policy, and ultimately by penetrating their own philosophy and framing their thinking. There are numerous examples of this in Swedish twentieth century history, beginning with the First World War when major companies gave their support to democracy. . .[to] the unlikely spectacle at the 1985 general election of the Moderates laying claim to the legitimate inheritance of Per Albin Hansson and the People's Home
>
> (Tilton, 1988: 389, my translation)

THE SOCIAL CONSTRUCTION OF POLITICAL DOMINANCE: 'RESIDENCE POLICY'

As the political vehicle for a strategic alliance, the Social Democrats came to office in 1932 at a time when Sweden was still only in the early stages of urbanisation. The ability to combine the emergent hegemony in society with political dominance through the state therefore took place in time to influence the direction in which the urbanisation process could develop. That influence did not make itself felt until after the Second World War when many of the plans of the government could be implemented free from the restrictions of a war economy. The early postwar years were a period of rapid urbanisation as well as rapidly rising living standards. The 1960s, in particular, which saw exceptionally rapid economic expansion, in subsequent political mythology have come to be known as 'the record years' (*rekordåren*) and have contributed significantly to the aura of political success of the social democrats.

It has been argued in Chapter 7 that an essential component of a divergence thesis is the way in which Swedish urban form has developed, with its predominance of flats, highly developed collective transport and other public services, and high levels of female wage-labour. The key question that must be asked in this context is to what extent can Sweden's distinctive urban form – and the pattern of residence that underlies it – be ascribed to conscious political ideology and to the urban planning that derives from this? The following discussion will attempt to provide an initial and tentative answer to this remarkably neglected question in Swedish housing and urban research. In doing so, I propose to focus upon the concept of 'residence' as the interstice of household and dwelling in its locational context. I do not attempt an explanation of state influence on the urbanisation process as a whole, which is a task beyond the relatively narrowly defined question addressed here.

The starting point for this discussion must lie in an understanding of the particular organisation of the Swedish state. The Swedish state has been relatively highly centralised ever since at least the time of Gustav Vasa and the era of Sweden as a major European power. A close relationship between State and Church which, among other things, made possible the development of a sophisticated population register, was a major factor in facilitating the effective recruitment and replenishment of the regiments required for the war effort. This tradition of centralised power was at the same time complemented by a well developed and self-confident system of local government, in a country where communications were often difficult because of climate and the swampy forests that cover much of the land. The effect of this on the formation of the modern state was to combine strong central overall control with a high degree of local delegation. This, in combination with the strong tradition of efficient and detailed planning, provided the basis from which hegemony could be converted into political dominance with relative ease. Essentially, central government provides the guidelines and sets the general form, while local government – *Kommun* – decides the practical and specific form that will be taken.

It is a pity that Gramsci had nothing to say about the Swedish state. Gramsci's main concern, as we saw in Chapter 6, had been to counter the fixation by socialists on using state power for reform or revolution that had been the legacy of socialism from the experience of the Russian revolution. In the case of Sweden, we have a society with both a strong state and a well developed civil society. It is this special combination that helps to explain the specific path of urbanisation taken in Sweden.

In the following discussion, I propose to consider first the ways in which the general idea of the People's Home found expression in a number of dominant ideas which came to influence social planning, and then to consider two ways in which these dominant ideas took political expression in terms of public policy towards residence, both at the national and the local level.

THE PEOPLE'S HOME AND SOCIAL PLANNING

The 1930s were a period of rapid urbanisation combined with increases in housing construction that began to reduce housing shortages. The war years saw a continuation of urbanisation but with drastically reduced housing construction and hence growing housing shortages as a result of the priorities of war. This meant that after the war urban planning would figure centrally in building the People's Home. During the war years, while policy measures to implement the People's Home were marking time, a discussion group of radical intellectuals began to formulate the basis of the reform policies that would realise the People's Home, and in those discussions issues of urban form loomed inevitably large.[3]

Prominent in this group (referred to hereafter as 'the ginger group'), whose members later took leading roles in the major government commissions of inquiry, was Segerstedt, who in 1947 was appointed by Uppsala University as Sweden's first professor of sociology.[4] In a seminal paper published in 1939 he formulated the problem of urbanisation in terms of alienation resulting from the breakdown of primary groups, drawing directly on sociological theory in the tradition of Toennies and Cooley. This breakdown was a threat to the new political order in that political democracy in terms of voting rights and political consciousness rested on the existence of democratic individuals, able to support and implement desired policies. Moreover, the breakdown of primary groups could only be counteracted by the activities of the popular movements as long as they were themselves based on primary group relationships. But the popular movements had grown in scale into secondary groups and lost their ability to base solidarity and democracy on primary relationships. Segerstedt termed the effect of this breakdown of primary groups 'group homelessness' (*grupphemlöshet*). His use of the word 'home' with its associations with the People's Home was hardly coincidental. Segerstedt concluded that democracy was under threat through the creation of a mass society in which group homelessness produced passive individuals leading to dictatorship. Democracy could therefore only be upheld by collectivism: active co-operation based on primary groups (Franzén and Sandstedt, 1981: 55–9).

Various members of the ginger group took up this theme in different ways. For some it required the provision of adequate schools for the education of democratic citizens. For others – and here Alva and Gunnar Myrdal (particularly Alva) were very influential – it meant anchoring women in the labour force on equal terms with men. For the architects it meant that the planning of new urban areas had to be based on the reconstitution of primary groups, and from this a new type of planning theory emerged, termed 'neighbourhood planning' (*grannskapsplanering*).

Franzén and Sandstedt (1981: 59) argue that one reason for the rapid spread of neighbourhood planning theory was that architects constituted a small and well-integrated group, with several prominent practitioners also members of the ginger group, notably Zimdahl, Åkerman and Åhrén. Be that as it may, the theory drew much from anglo-saxon planning in terms of 'community centres', and the idea of satellite towns built around London as self-contained communities with all facilities provided: some centrally and others in local neighbourhoods. However, this took expression within the very specific context of Swedish political planning, and in particular the building of the People's Home. As such it was coloured with quite different ideological overtones. This contrast between British and Swedish neigh-bourhood planning mirrors the contrast between conservative and socialist mythologies of the People's Home.

Neighbourhood planning had a considerable effect on town planning, and the manner in which urbanisation took place in postwar Sweden reflects this. New suburbs were built which were based on the provision of collective facilities at the level of the block and the multi-dwelling build-ing, while larger aggregates were provided with local shopping centres and other sub-metropolitan facilities (Franzén and Sandstedt, 1981: 71–82).

Franzén and Sandstedt (1981) depict the process of Swedish urbanisa-tion in terms of Marxist theory, drawing heavily on Habermas and arguing that the manner of urbanisation in Sweden took expression in a compromise between the interests of capital and labour (Franzén and Sandstedt, 1981: 278). The outcome in terms of urban form was unique – though how unique can only be determined by comparative studies with urban form in other countries – and 'the result of a specific combination of a whole number of factors' (Franzén and Sandstedt, 1981: 311; my translation).

It is precisely these factors which I wish to consider further here, and the clues are to be found in the detailed evidence that Franzén and Sandstedt themselves present. There are many aspects of this that could be examined more closely in terms of the effect of People's Home ideology on urban form. Here I want to focus on just two of these which mark out Swedish suburbanisation from similar processes in other countries, and particularly in English-speaking countries. They are the provision of collective facilities at the neighbourhood and multi-dwelling building levels and the town planning conflict between high dwelling-space standards and urban sprawl.

The Collective Housing Committee: social planning and the home

The basis for central government action in Sweden lies in the careful investigation and planning that underlies the passing of acts and the imple-mentation of administrative measures. These commissions of public inquiry (SOU, or *Statens offentliga utredningar*) have a long tradition in Sweden, but became of even greater significance in the construction of the People's Home. An early concern of the social democratic government was the improvement of official statistics, and one of their first commissions of public inquiry concerned the improvement of housing statistics (SOU, 1933). Since the Second World War the number of inquiries has increased, as has the complexity and depth of the research behind them. They cover an astonishingly wide range of topics from inheritance tax to government support for the Swedish film industry and from education for artists to the gender classification of trans-sexuality.

One of the fundamental principles underlying the People's Home was that of equality between citizens, and particularly between men and women. This gave rise to considerable debate, in which Alva Myrdal was particularly prominent, over how neighbourhoods could be organised in the rapidly expanding towns and cities in such a way that did not result in women becoming trapped in remote suburban houses, and which also contributed to sustaining the sense of community and social consciousness expressed through the popular movements.

The need to provide collective facilities in neighbourhoods was a well established government commitment by the late 1930s. However, at this time and until the 1950s, the principal concern was not to enable women to sell their labour but to free up some of their time so that they could participate in the democratic process as equal citizens along with men (Franzén and Sandstedt, 1981: 217–18). At this time, then, the role of women was not envisaged as changing and this is reflected in the fact that it was only in the 1960s that the housewife role ceased to be the norm. By then, the acute labour shortages that were emerging in Sweden provided extra incentive for the changed role of women.

By way of example, I wish to consider the more detailed proposals set out by one commission in the mid-1950s. In 1945 a Commission of Investigation into Population recommended an investigation into the provision of collective facilities in neighbourhoods. Another Commission, into Home and Family issues, which reported in 1947 on technical and planning questions concerning the rationalisation of housework, recommended a Commission report on the planning and provision of guidelines for collective facilities. In 1948 the Minister of State for Social Affairs, Gustav Möller, therefore assembled a group of seven experts – one of whom was the architect Åkerman, member of the ginger group – to conduct a series of investigations concerning the need for collective facilities in housing areas.

During the following eight years this Commission produced a number of interim reports on special topics: homehelp (SOU, 1952), collective housing forms (SOU, 1954), the social organisation of laundering (SOU, 1955a), the provision of public meeting locales (SOU, 1955b), and a final report on the home and societal planning (SOU, 1956). These reports were in the best traditions of Swedish planning. That on laundering, for example, presented 368 pages of statistical, pictorial, and organisational data on four different forms of laundering (in each home separately, joint facilities for each multi-family dwelling unit or group of houses, neighbourhood launderettes, and neighbourhood laundries), together with evaluation and conclusions. The Committee concluded that the most effective form was that of the provision of facilities for each multi-family building or group of dwellings but that these would need to be of a higher technical standard than hitherto (SOU, 1955a: 364). It is interesting to note that this form of collective laundry provision did in fact become the dominant form, at least in blocks of flats, and remains so to this day.

The reports covered not only the provision of facilities in the form of buildings and equipment but service support and its organisational form, for example in terms of homehelp and childcare facilities. These covered six major categories (SOU, 1956: 14):

1 Rationalising the provision of food and meals: shops, various forms of collective meals, school meals, restaurants, etc.
2 Rationalising care of clothes and laundering.
3 Collective homehelp, including special categories of care.
4 Childcare and youth facilities, including daycare and playschools, club facilities, parks, playgrounds and sports facilities.
5 General meeting places for different age groups and interests.
6 Technical equipment common to both larger and smaller areas besides water drainage and electricity, such as district heating, bath and sauna facilities, collective refrigeration and freezer facilities, polishing machines and other expensive, technical, labour saving devices.

The ideological motivation behind this series of reports was clearly two-fold: first and foremost to free women from much domestic drudgery, and secondly to facilitate the reconstruction in new housing areas of collective social life in which individuals would feel a sense of belonging and solidarity.

The first of these was unambiguously the major concern. The final report of the Commission (SOU, 1956) spells out in some detail the reasoning behind their work. It argued that 'a new spiritual atmosphere' has emerged with the process of industrialisation, based on rationalism, individualism, and respect for human values (SOU, 1956: 29–32). Central to this process is equality between members of the same family – and here the secure grounding of the People's Home in the family home is quite clear – especially between men and women, as a result of the growing need for female labour and the concern that this should not result in 'double labour' for a woman through running the home and selling her labour to an employer (SOU, 1956: 33–4).[5]

The solution to this problem is seen in terms of reducing the burden of home work by a combination of the introduction of technical equipment and, more important, the provision of collective facilities in the form of labour and time saving equipment and childcare facilities. The Commission made it clear that the primary consideration was the ideological one of equality for women as a key component of the People's Home. The societal benefits in terms of efficiency, public expenditure savings and the smooth functioning of capitalism was only a secondary consideration (SOU, 1956: 35–6).

The other aim – creating a sense of belonging and the facilitating of primary relationships – was also important. As the Commission put it:

> Meeting place facilities of various kinds have the overall purpose – particularly for the benefit of societies and associations – of providing locales for the pursuit of the kind of citizen activities that are the precondition for a democratic society, at the same time as they make it possible to remove additional space demands made by spare time activities on normally overcrowded housing.
>
> (SOU, 1956: 23; my translation)

Here we find the expression of the need to nurture primary groups in housing areas in order to produce democratic citizens, particularly for the nurturing of popular movements. But there is an additional argument: easing overcrowding in the home by removing free time activities from the home and locating them in specially provided collective facilities. This provides an interesting clue to the second example below, and will be taken up again in that context.

The report discusses in some detail the ways in which the planning of collective facilities should be built into the organisation of the agency form responsible for implementing local policy, namely the board, or delegation (*nämnd*, literally 'named') both at the different levels (*kommun*, building and development etc) and the numerous specialist boards such as childcare, social care, and homehelp (SOU, 1956: 163–8). They also propose the setting up of a co-ordinating agency for housing collective facilities (SOU, 1956: 171–3). Concerning national measures, they propose research and education, better planning guidelines, and support in the form of loans and grants. They also propose the establishment of a special board for collective facilities within the National Housing Council (*bostadsstyrelsen*), the national organ which is responsible for fleshing out legislative and administrative policy (SOU, 1956: 174–86). Such a board would provide expert advice on the provision of collective facilities and 'contribute towards inculcating Swedish housing policy with the values we have sought to give expression to here' (SOU, 1956: 185; my translation).

The extent to which collective facilities have been provided in Swedish neighbourhoods is impressive – that this occurred even before the 1956 report is evident from the numerous examples cited in it – and quite clearly distinguishes Swedish urban areas from those of English-speaking countries. New Australian housing areas in particular have been heavily criticised for providing no initial collective facilities – not even schools or shops – and for leaving the new residents to make do, sometimes for years, until public agencies and private companies begin to move in and provide basic services (see Kemeny, 1981: 47). And even then there is no comparison with the range and extent of facilities provided for in Swedish planning law, such as those described above. At the same time it is important to distinguish between the way in which ideology has determined urban form and the degree to which the implementation of such ideological aims has produced the expected results. Swedish neighbourhoods, by providing high levels of collective facilities, would appear to have contributed significantly to more wage-labour oriented female roles. However, it would appear equally clear that the expectation that collective facilities would foster solidarity and democratic participation has been a monumental failure (for a discussion see Popenoe, 1977). This point will be returned to at the end of the chapter in the context of a more general assessment of the success of the establishment of a collectivist hegemony in Sweden.

Town planning: dwelling type and collective urban form

The extent to which dwelling type is linked to ideological preferences for collective urban form is much less easy to determine than the ideological arguments behind the provision of collective facilities. Government commissions of investigation into housing concentrate on the issues of housing shortage as measured in numbers of dwellings, issues of housing standards as measured in facilities (such as hot water, central heating and inside toilet) and overcrowding (as measured in persons per square metre, room, etc.). They handle dwelling type only in so far as there are cost differences between them when calculating subsidies (see, for example, SOU, 1945, 1974). The calculations made of these are based on detailed costings which take into account the need to raise minimum housing standards. They are also based on more all-embracing planning principles – for example on the provision of a wide range of collective facilities – than is normally the case in English-speaking countries.

The net result is usually that, in stark contrast to conclusions drawn in English-speaking countries, flats are concluded to be cheaper than houses. The cultural differences in terms of the conclusions drawn concerning the relative cost of houses and flats have been discussed elsewhere (Kemeny, 1981: 46–7). Here it need only be noted that the ideological preference for flats in Sweden among both government housing advisors and town planners is not based directly on 'ideology' but on economic considerations which already have built into them certain social principles concerning the need for overall planning, for the provision of high levels of collective services in housing areas, and for encouraging high minimum housing standards by tying subsidy eligibility to ambitious minimum norms. Precisely the same ideological dynamic takes place in English-speaking countries, except that the privatist dominant ideology leads to diametrically opposite conclusions being drawn. The practical effect of this in both countries has been to remove the issue of dwelling type from public debate, except in the special case of high-rise housing.

The theoretical implications of this are interesting because the ideology is built into the economic calculations that form the basis of housing policy, and so policy appears on the face of it to be 'rational' and based on strict 'economic logic'. What we have here is a rather interesting example of the way in which the ground rules governing decision-making by policy-makers are determined by underlying ideological influences which themselves constitute part of the negotiated order. They are never questioned because they remain part of a hidden agenda, one that is pre-determined by the social construction of the economic statistics that constitute the apparent moral neutrality of the raw material upon which policy is based.

There have been attempts to redefine the moral basis of official Swedish housing statistics in recent years, by showing how the cost of housing can be decreased simply by lowering or abolishing the construction norms that are the basis of government subsidy eligibility. The problem for such advocacy is that there is considerable agreement that some norms are needed and it then becomes a matter of judgement of which norms should be kept and which abolished. This considerably weakens the criticism and makes the existing moral basis of the negotiated order seem 'reasonable' and so harder to challenge.

The lack of ideology in national housing policy over dwelling type is matched by an absence of apparent central government concern for the content of town planning. The reason for this is somewhat different, since planning is essentially a local government responsibility. Government commissions of investigation into planning are therefore limited in general to setting the framework within which local planning is to take place: for example, in the legislation to provide guidelines for the production of town plans by local government, which was passed in the war years.

However, an ideological preference for dwelling type does surface occasionally. One particularly clear example was in the appendix to the 1945 housing policy investigation, entitled 'Systematically planned Society-building' (*'Ett planmässigt samhällsbyggande"*: SOU, 1945: 580–640). This was written by Århén, one of the five members of the Commission and also one of the architect members of the ginger group. Århén argues in this appendix for the concept of neighbourhood planning, basing his arguments on ideas in sociology concerning group membership and the need for primary groups. He is critical of the poor provision of collective facilities in urban areas as a basis for the existence of primary group membership. He argues that collective urban facilities are needed to provide the basis for the 'development of a democratic type of person, characterised by active self-motivation (*självverksamhet*), initiative and co-operation' (SOU, 1945: 594).

In a section on 'urban sprawl' (SOU, 1945: 596–601), he strongly criticises the unplanned spread of detached housing on the edges of cities as a reflection of the modern antisocial and disoriented individuals who are lacking cultural collective interests, are without a societal focus point to orientate themselves towards, have limited social contacts, and have an acquired habit of avoiding having to take account of neighbours (SOU, 1945: 597). The economic argument against urban sprawl also surfaces in that the costs of providing collective facilities for low densities would be enormous, and therefore more dense construction must be preferred (SOU, 1945: 599). However, for Århén, although there are economic arguments against urban sprawl which he details, the social arguments are the more important:

> In the final analysis the social reasons are decisive. According to the goals set out in the chapter 'Sociology and society-building', measures should be directed towards creating the kind of society which encourages the emergence of individuals who are socially minded and able to work effectively together. If one wanted the opposite characteristics one would recommend urban sprawl, which offers neither a natural collective context nor social interests in common.
>
> (SOU, 1945: 601; my translation)

Århén concludes by stressing the urgency of the situation, and hence the need for immediate and decisive measures, since urbanisation will have been more or less completed by the 1970s or 1980s:

> If one wishes to implement systematically planned society-building then it has to happen soon, otherwise it will be too late. Our towns and other urban areas began with industrialisation to expand rapidly in the 1870s and 1880s. It was then that their present form was decided. By the 1970s or 1980s our urban areas will be largely fully formed. By that time even the renewal of older areas will have been completed. We therefore have barely more than the last two or three decades left of the last 100 years of urbanisation to right the major errors of the last seven or eight decades. Much will depend on the next ten years.
>
> (SOU, 1945: 638–9)

Åhrén's polemic against urban sprawl is only indirectly a critique of detached housing, and nowhere does he raise the issue of dwelling type explicitly or argue for flats or terraced housing in preference to detached housing. Nevertheless, the very fact that he elides the distinction between dwelling type and urban form does create 'guilt by association'. It is likely that the very strength of the neighbourhood movement and its vocal expression encouraged an implicit preference among planners for flats over houses, a preference that was given empirical support in the housing finance statistics and the loan conditions that underlay new build and renovation.

But while political discourse at government level over dwelling type remained muted, there was considerable argument taking place at other levels. Rådberg (1988) has discussed the different ideological positions among architects and planners in considerable detail. An important element in the preference for flats over houses was the fear that if the proportion of houses built was increased the better off would buy them and abandon flats, leaving the stock of flats as 'category housing' for the worst off and thereby creating class differences in residence. An important goal among many planners was therefore to facilitate the production of flats as cheaply as possible and not to encourage house building overmuch. As Rådberg puts it:

> The postwar publicly-managed multi-family dwelling was seen as confirmation of the fact that society was on the way to realising an important political goal of attaining the People's Home – a society with class differences broken down in which different social classes lived side by side. This easily led to the belief that an increase in house building must be a symptom of ideological failure in this respect. Mistrust of increased house building would have been heightened when it became clear that the detached house suburbs around the cities manifested bourgeois electoral sympathies.
>
> (Rådberg, 1988: 338; my translation)

Rådberg (1988: 338) also points out that there was a strong pro-owner occupier element in social democratic thinking, and here we have a possible clue to the attempt to achieve a 'tenure-neutral' housing policy in Sweden (Lundqvist, 1988), since the ambiguity and lack of a strong line on this issue made tenure-neutrality the default option that state agencies could and did implement both nationally and at a number of local levels.

This only partially formulated preference for high densities is well illustrated in the practical implementation of national policy at the local level. There were wide variations in local government implementations of high-density planning at different times. In the early decades of this century, towns such as Västerås and Kalmar built terraced housing for working class families extensively (Rådberg, 1988: 190, 215). The ambiguity towards dwelling type is well illustrated in Stockholm, in the city's General Plan which was produced at the end of the war. It came in two parts: guidelines published in 1945 and the main plan itself in 1952.

The production of a city plan was the direct result of legislation passed during the war to increase the effectiveness of urban planning as an integral part of building the People's Home. At this time Stockholm's population had passed 800,000, of which only 150,000 were located in the suburbs, primarily *kommuns* to the northwest (Sundbyberg, Solna and Hässelby). Predictions of population growth suggested that it would pass one million before 1960 and level out at around 1.1 to 1.3 millions. Stockholm therefore stood at this time on the threshold of a major expansion, and the plan therefore had the task of determining the urban form that the city would ultimately adopt.

There were clearly a number of options and three of these, drawn from international experience, were set out in the guidelines by way of illustration. One was the Soviet model, based on transhumance, in which each family would have a central rental flat for the winter and a dacha in the country for the summer. Another was the British and Dutch model involving relatively high densities but based on terraced and detached houses served by well developed public transport. The third was the Los Angeles model, epitomising urban sprawl and private car dependency, with a population less than double that of Stockholm but covering an area more than eight times larger (Stockholm, 1945: 29–30).

It was observed that decisions needed to be made concerning the balance between high housing space standards - which were a major housing policy aim - and long commuting distances. The conclusion reached was that Stockholm's expansion should take place through the construction of suburbs in the form of satellite towns (*drabantstäder*), each provided with its own collective services and linked by public communications to the city (Stockholm, 1945: 35–6).

Chapter 4 of the report handled the question of choice between different types of housing. It was pointed out that the overwhelming majority of the population lived in rental housing (this being equivalent to the dwelling type of flat for all intents and purposes), with only 17 per cent of the population living in one- or two-family houses. It concluded that a substantial expansion of detached housing provision was not desirable for three reasons: the traffic problems created by urban sprawl, the high cost of providing infrastructure (street and house piping and wiring), and the poorer level of collective services that can be provided.

This final factor was decisive because collective facilities are important in easing the burden of housework, a necessary contribution to facilitating the commodification of female labour (Stockholm, 1945: 46–8). This led the report to argue for heavier investment in collective facilities for traditional housing and for collective housing (flats with facilities such as joint kitchen and lounges, which are normally provided privately, in addition to childcare, meeting places, laundry and other collective provision).

On the basis of this, the report concluded that the dominant dwelling types should be three-storey rental flats for higher densities and terraced houses – drawing on the British experience – instead of detached houses for owner occupation. It also suggested that the plan should take into account the likelihood that holiday homes (*sommarstugor*) would become more common, implying clearly that, of the three possible models to choose between – Soviet high density, Dutch or British medium density, and American low density – the proposed pattern of residence would follow most closely the Soviet model.

The guidelines proposed that satellite towns of some 14,000 people would be the optimum size for suburbs, each with an inner area of flats for some 6,000 people adjoining the shopping centre and other public facilities and an outer ring of terraced houses for 8,000 people (Stockholm, 1945: 57). The guidelines then went on to discuss slum clearance, transport and communications, and other aspects of town planning, which do not concern us directly here.

The interesting point to note is the extent to which high density urban form featured as a major characteristic of town planning. The urban form that actually resulted from this planning differed somewhat in that terraced housing never came to replace the detached house in the manner that was proposed. On the other hand, the numbers of flats appear to have greatly exceeded those proposed. It is striking to note that while the proportion of houses (as against flats) was 17 per cent in 1945, this had increased to only 29 per cent by the early 1970s (Kemeny, 1981: 44). Yet the guidelines proposed that over half of new build would comprise terraced houses. This should, in theory, have increased the percentage of houses to around 35 per cent, but resulting in higher population densities than have in fact been achieved.

The reasons for this would seem to be the failure of housing policy to encourage the construction of terraced rental housing. Swedish tenure remains to this day highly dwelling-type determined, with rental flats and owner occupied detached housing. Terraced housing has increased in recent years but remains a minority dwelling type, mainly in the co-operative housing sector. This disjuncture between planning and outcome constitutes an important link between ideology in its planning phase and what actually happens on the ground and this subject remains to be researched adequately.

One reason for this disjuncture must be the failure on the part of the ideologues to address head on the issue of dwelling type and tenure, resulting in a lack of clarity over these issues in terms of implementation. The ideological preference for a particular dwelling type appears to have remained implicit and largely part of a hidden agenda, and not become part of wider political and social debates. This was in stark contrast to the issue of collective provision of facilities to create democratic individuals and to facilitate women moving into the labour market, which had its ideological roots in the People's Home and its intellectual basis in the deliberations and polemics of the ginger group. However, it may be that more detailed research can uncover the existence of an ideological debate in which the ground rules favouring flat construction became explicitly formulated into the negotiated order of housing and planning policy. But if this is so, it remains an indisputable fact that this issue never became a central one in the conversion of hegemony into political dominance in the way that was true of, for example, the need for collective facilities or the emphasis on public rental housing. It is precisely such ideological lacunae that provide the opportunity – or rather the power vacuum – for state apparatuses to exercise autonomous discretion, often with unpredictable results.

CONCLUSIONS

Swedish housing policy must be understood in the context of the wider aims of the development of the welfare state. Central to this has been the creation of urban areas which are designed to minimise the amount of female domestic labour so as to encourage female wage-labour, and to foster social solidarity as a means of defending the People's Home against dictatorship and reaction. Both of these aims also constituted elements in the broader historical bloc of the popular movements in terms of aims of equality, particularly between men and women, and the fostering of democracy.

In attempting to implement such aims, the intimate relationship between housing and urban planning is highlighted. Indeed, it is ultimately impossible to separate out issues of housing from urban planning. If we are to understand Swedish housing policy it is therefore necessary to see the relationship between housing and wider social concerns: the role of women; attitudes towards democracy; models of social participation and the role of the individual; urban form, and so on. Housing then becomes one major element in a complex interrelationship, constituting a basic 'key' to unlock differences between societies at a fundamental level of social organisation that is more usefully understood in terms of patterns of residence than in narrow housing issues.

The brief and selective discussion of issues of Swedish residence policy presented above shows how the conceptual framework outlined in the previous two chapters is useful for understanding the special characteristics of Swedish society. Despite the limited context, it is possible to trace the social construction of hegemony, and the manner in which this took expression in public policy as a means of reinforcing hegemony with political dominance. State action was vital to concretise aspects of hegemony in terms of the legal framework of planning and building, and to implement ideologically motivated goals which had little or nothing to do with the logic of capital. This contributed in a major way to the distinctiveness of Swedish urban form and the collectivisation of social life.

But how succesful was this policy in bringing about fundamental changes in Swedish social structure? It is necessary to conclude that the results have been mixed. The policies have succeeded to a considerable degree in creating social structures, particularly in urban areas, that support collectivism in life styles and transport and encourage female wage-labour. But at the same time there is a limit to the extent to which such policies and planning imposed through state agencies can implement far reaching collectivism in social structures. As discussed in Chapter 7, socialist strategy has been to develop hegemony only as a stepping stone to attaining political dominance, and has used political dominance as a means of changing the state in the belief that this would be sufficient.

The result has been the establishment of collectivism in terms of state provided welfare and in the case of Sweden also in more extensive social planning for collectivism. But there has been no sustained attempt to use hegemony to create collectivised social structures. Indeed, state provision of collective services has to a considerable extent been counterproductive for the social construction of deep collectivism by making grass-roots co-operation unnecessary in many key areas. The result has been the impoverishment of populist movements and the oligarchisation of their organisations and the incorporation of these into a hegemonic corporatist state (see Figure 6.1). The collectivist ideology has lived on, long after one of its main purposes – to foster grass-roots radicalism – had been abandoned. The massive 'million programme', in which one million dwellings were built in the decade 1965/74, thus perpetuated the form of collectivist residential structure but all talk of its fostering democratic individuals had disappeared. The collectivistic ideology remains alive to this day, most recently manifested in the Commission of Investigation into Cities (SOU, 1990).

Swedish social structure is therefore characterised by extensive privatism. What distinguishes Swedish social structure from that in, say, anglo-saxon societies is that political dominance, in conjunction with a powerful tradition of state intervention and control, has been used to intervene in social structures to create a considerable degree of collectivism – particularly in terms of residence – which provides important social structural support for the continued existence of the welfare state.

The evidence of Swedish housing and urban policy suggests that it differs very markedly from that in countries such as Australia, the USA and Britain. In these countries, and particularly in Australia which lacked an old stock of high density housing, no systematic and prolonged attempt was made to counteract low density car-dependent housing provision, nor to encourage female wage-labour through urban restructuring. In Britain urbanisation took place over a very long time and with a variety of different political influences upon it. Most of the time there was a generally centrist – for most of the postwar period called 'Butskellite' – dominance without radical left or right wing influence ever dominating for long. The 'passive revolution' of Thatcherism came too late to affect urban form, which was already established by the 1970s,[6] but in a sense it would have been unnecessary. This is because residence has always constituted an arena of 'non-decisions', in which an unchallenged privatist hegemony has reigned supreme. Where labour parties have held office, the state has been used to boost public welfare or to nationalise private industry in narrowly conceived statist terms, but never to penetrate social structure with collectivism. The fact that this was done in Sweden with enough sucess to give Sweden a distinctive social structure has almost certainly been a decisive factor in the creation and continuance of a stable welfare state.

Both Australia and Britain would be interesting examples of dominant ideologies in housing policy. In both countries the labour movements have willy-nilly adopted the dominant mode of discourse of a right wing hegemony as a result of a long period of labourism class ideology. In Britain, the Labour Party has adopted the rhetoric of council house sales and 'market socialism'. In Australia, Labor ministers talk patronisingly about 'welfare housing', meaning not, of course, owner occupation with its massive subsidies but public renting where subsidies are low or even in some cases negative. These are all clear signs of incorporation of the opposition into a dominant mode of discourse and the absence of an alternative with which to challenge it. Such is the power of ideology to define reality.

Part IV

Toward a theoretically anchored sociology of housing

9   Residence and social structure

INTRODUCTION

The failure to develop an epistemology of housing lies at the root of the ambiguous position of housing research in the social sciences and the neglect of housing as a dimension of social structure. Central to this is the problematic and unclear substantive focus of housing research, deriving from tensions between the concepts of household and dwelling that in themselves are not resolvable. However, it should be possible to go beyond this by developing a higher level integrating concept that would provide a better basis for problem formulation and would lead analysis naturally towards a conception of housing more clearly related to social structure. This would result in a substantive focus that is more in tune with a theoretically explicated housing studies well anchored within the social sciences. The primary task in this chapter is to begin to address the question of what is the most effective conceptual basis and substantive focus of social science housing research.

THE SALIENCE OF HOUSING IN SOCIAL STRUCTURE: THE CONCEPT OF EMBEDDEDNESS

Housing research focuses on housing as a dimension of society in its own right, in abstraction from other, 'non-housing' social, economic, and political dimensions. The rapid institutionalisation of housing studies in higher education and research is likely to reinforce the tendency to view housing as an abstract subject. A key problem is therefore to conceptualise the integration of housing in social structure in such a way that its salience is accurately reflected, with housing neither being elevated to an abstract field nor relegated to a minor bricks-and-mortar issue with little social importance.

A particularly fruitful approach to this problem has been developed by Granovetter (1985) in his concept of 'embeddedness'. Using the economy as an example, he argues that explanations of how markets work too easily fall into the opposing errors of the undersocialised and oversocialised conceptions of human action. In the former, the market is seen as largely governed by its own rules of rationality with social and cultural factors as extraneous non-rational influences that gradually disappear as perfect competition develops. He terms this an 'undersocialised' conception of human action in that it sees the market as something that is non-social and apart from social structures. In the latter, the market is seen as dominated by structures of legitimation and power. Following Wrong (1961), he defines this as an 'oversocialised' conception of human action in which the economy is determined by social relations and lacks autonomy.

Granovetter argues against both of these views, claiming that economies can be better understood as being *embedded* in social structures. By this he means that behaviour and institutions – in markets or indeed in any form of social organisation - must be understood as being located in broader social networks of sociability, approval, status, and power. These are normally neither insignificant and marginal nor dominant and determining in their effects on the workings of markets. Rather, they constitute the grounds upon which institutions such as the market rest and without reference to which they cannot be properly understood.

The trick in conceptualising embeddedness is to reflect the degree of salience that has the greatest explanatory power. For example, the undersocialised view of the market can be contrasted to the oversocialised view of the state in early postwar political analysis, in which the state was discounted as an autonomous institution in analyses of power (see Chapter 3). Appropriate salience is achieved by stressing the ways in which the state as a social institution can and does act autonomously of vested interests in the wider society. The net result is to achieve a degree of 'disembedding' of state institutions from the analysis of societal power, thereby heightening the salience of the state.

Both of these tendencies – the oversocialised and undersocialised approaches – can be seen at work in the way housing is treated in relation to social structure. This takes the form of a special kind of 'housing rationality' or logic. According to this, housing is either seen as so embedded in social structure as to be deemed inconsequential, and so indirectly handled as a result of analysing other aspects of social structure such as households, the home, community, etc., or it is elevated into a specialism in its own right in which it is abstracted from social structure.

The former represents the oversocialised conception of housing in that outside of housing research housing loses salience to the point of invisibility. Just as the salience of the state was reduced to inconsequentiality when political sociology focused upon the social bases of power at the expense of 'narrow' political institutions, so housing, as a central defining fact of the everyday life of household members and as a major dimension of society, is all too easily defined as a purely physical – and hence sociologically residual – factor. Just as the state is defined as a black box through which vested interests express their interests, so housing becomes a physical medium through which households engage in social action. Housing is thereby seen as being so deeply embedded in social structures that it becomes inconsequential and disappears for the purposes of analysis.

The latter represents the undersocialised conception of housing in that it allows housing to be researched in abstraction from other dimensions of social structure in much the same way as classical economists argued was the case for the market. This is an approach which remains influential today. Indeed, the very existence and success of a 'housing studies' is predicated, albeit in an unexplicated manner, on the disembeddedness of housing from social structure and its undersocialised nature. Attempts to overcome the isolation of a housing dimension by linking it to another dimension, such as employment or education, are inadequate compromises that serve only to underline the non-embeddedness of housing in social structure.

Upon what is this internal housing rationality based, and how is it constituted? To understand this it is necessary to return to the basic concepts that underlie housing research and the problematic nature of their formulation.

RESIDENCE AS A SOCIO-SPATIAL FOCUS

One of the consequences of the unresolved tension between the concepts of household and dwelling that has been a recurring theme in this book is a shifting and unclear substantive focus. This tension derives from the fact that the concept of a household is primarily a social one while that of a dwelling is primarily physical or spatial. This has created an underlying ambivalence in housing research. Housing is seen in the final analysis as the study of physical dwellings in terms of their production, management and disposition. This sometimes involves the study of dwellings as purely physical objects, as, for example, in the study of a stock of public rental houses and its loan and rent structures. But at other times it involves the study of households, notably over issues of affordability and in studies of residential mobility, at which point housing research becomes clearly social in focus with little or no interest in the dwellings themselves. More commonly, analysis swings unthinkingly – and sometimes with breath-taking speed – between physical and social dimensions as different aspects of the tensions inherent in the household–dwelling relationship intrude upon the awareness of the researcher at different points in the research. This, I want to argue, constitutes a fundamental weakness in the very grounds of housing research, which lies at the root of the undersocialised conception of housing that dominates housing studies and the over-socialised conception of housing that predominates in sociology.

The tension between the social dimension of housing in terms of house-holds and the physical dimension in terms of dwellings suggests that a focus for housing studies within the social sciences might well be found in terms of a combination of the two. How might this be specified? The social dimension is clear enough, but the physical dimension and its relationship to social factors needs to be spelled out.

Following the approach foreshadowed in Gregory and Urry (1985), it is fruitful to conceive of the household–dwelling relationship in terms of an integrated socio-spatial dimension. To this end, I want to argue here that space is the salient characteristic of the physical dimension of dwellings in relation to social factors. This comprises two levels: the internal spatial organisation of dwellings and their social use on the one hand and the spatial organisation of dwellings within the locality on the other. The former relates to such questions as overcrowding, multi-occupation, and relationships within and between households occupying the same dwelling. It therefore directly addresses the ambiguity of the relationship between household and dwelling. The latter relates to the impact of dwelling type and densities on neighbourhood and urban form, and, as we shall see, has wide ramifications for social structure.

These socio-spatial relationships centring on housing can best be des-cribed in terms of 'residence', which encompasses both internal dwelling and external locality factors. Residence in the sense of residing in a dwell-ing in a particular locality therefore focuses on the socio-spatial significance of housing. It includes all of the traditional housing concerns, such as financing, constructing, managing, and disposition, but in addition it includes the spatial impact of the dwelling itself. In the first instance, this is on the household(s) inhabiting it and, in the second instance, in com-bination with household characteristics, on urban form and social structure in general. Most important, it focuses on the *interaction* between household and dwelling and their combined effect. Finally, residence places the indi-vidual – more accurately the household – at the centre of analysis by focusing on the act of residing and its socio-spatial implications. To this extent the approach is basically social and has a clearly sociological orientation.

Conceptualising the subject matter of housing in terms of residence suggests that housing can be analytically broken down into a threefold distinction of levels of increasing scope of social structure, namely household (including, for example, its composition, stage of family cycle, socio-economic status), dwelling (including its type, size, conditions and facilities) and residence (including shops, laundries, take-aways, collective transport, etc.). At the centre lies the relationship between household and dwelling. Beyond this lies the relationship of the household-in-dwelling to local society: town, suburb, village, rural area. Particularly important in this context is access to work, schools, shops and other services and facilities. At a more structural level is the relationship of the household-in-dwelling to the larger institutions and organisations of society that impinge directly and indirectly on issues of residence: housing finance institutions, central and local government, etc. We may consider these in turn.

The household–dwelling relationship

At a very basic level, housing provides the space which frames and often defines many of the activities that constitute primary relationships in the home. As we have indicated in Chapter 1, Saunders and Williams (1988) have argued that the home has been neglected as a research area, and they have made out a case for developing it as a substantive focus in its own right, separately from housing issues. As they put it:

> Our focus is on the home *rather than housing or the dwelling itself* and this opens up a range of crucial and related issues concerning, for example, household structures and relationships, gender relations, property rights, questions of status, privacy and autonomy, and so on. Perhaps surprisingly these issues (and their links to broader housing concerns) and the home as a whole have been somewhat neglected within the literatures on the sociology and political economy of housing, although a focus on the home does have a long history in other disciplines such as social anthropology.
>
> (Saunders and Williams, 1988: 81; emphasis added)

Saunders and Williams are on the right track in trying to infuse housing issues with those of the households that inhabit houses. However, there are problems with this approach, and it has engendered some debate (Dickens, 1989; Gurney, 1990; Somerville, 1988). For the purposes of this discussion, however, rather different points must be noted.

In the first place, Saunders and Williams do not provide an epistemological basis for this shift of focus. They do not attempt to develop a theoretically anchored set of justifications for focusing on the home.

Gurney (1990) argues that Saunders and Williams define the home in narrow *physical* terms, as the inside of a dwelling, drawing on Giddens' concept of a locale (which he in turn unwittingly had borrowed from negotiated order theory: see Chapter 6) to define the home as a place circumscribed by the boundaries of a dwelling within which identities and meanings are formed. This, while convincing, is a somewhat restricted interpretation. In practice, Saunders and Williams are interested in the home in the broader sense of a set of meanings and identities central to understanding modern society. But in so far as this remains theoretically unexplicated it opens them up to criticisms of the sort that Gurney makes.

The net effect of this is that the focus on the home remains only a change of subject - or focus - from dwellings to households, backed by a rich literature to justify its usefulness. It might be added that Saunders and Williams' argument explicitly comprises part of a wider concern to focus upon consumption rather than provision issues. This is a complicating factor in their concern to put the home at the top of a research agenda which has more to do with substantive issues than epistemological ones, and more to do with consumption issues than housing ones. Nor does the shift in focus away from dwellings and towards the households that inhabit them resolve the conceptual ambiguities surrounding the substantive focus of housing studies. This is exemplified by the way in which the words 'house' and 'home' are often used interchangeably or are closely coupled as in 'house-and-home'.

The approach of Saunders and Williams is a good example of the application of the oversocialised conception of housing in which the salience of housing is reduced by embedding housing into the relationship between household and home to such an extent that the dwelling loses significance almost completely. Attention is focused instead upon the home in which the dwelling becomes a mere arena for the playing out of inter-action, both within the household and between household members and the outside world. But the home is strongly influenced – even formed – by the socio-spatial relationship between household and dwelling. The home cannot be understood except as a product of the social organisation of the household in relation to the dwelling as a spatial reflection of that organisation and the limitations that this places on, and the possibilities that it opens up for, household members' activities and relationships. Therefore, the neglect of that relationship must necessarily impoverish any understanding of the home.

The household–dwelling relationship is an important building block in a residence-based perspective on housing. But it cannot be understood in isolation from more general issues. The choice of a dwelling by a household is intimately connected to the choice of dwelling location, and physical access to social resources. Over and above this individual-centred perspective it is necessary to take into account wider social factors which determine patterns of residence.

Households-in-dwellings and the wider society

Households rarely choose a dwelling for its internal attributes alone. Characteristically crucial in any choice of dwelling is its local setting: its spatial relationship to other dwellings, the street, and local area, and accessibility from the dwelling to workplaces, educational institutions, shops, friends and kin, recreational facilities, and the wider web of dependencies that determine life style and well being. In short, the location of the dwelling constitutes one of the key elements – if not the key element – in the social integration of individuals into society. It determines the manner in which individuals will be knitted into the various relationships that constitute their everyday lives and work: whether, for example, it will be necessary to commute to work, whether a car will be needed (or even two or more cars), and how much contact there will be with relatives and neighbours (which may affect the extent of mutual aid over childminding and care of the sick and elderly, etc.)

In all of this, the spatial impact of dwelling type on the nature of local society, as discussed in Chapter 7, is well known, yet its significance is remarkably poorly appreciated. Residential densities at the block level and beyond are largely determined by the dominant dwelling type that has been built. This in turn determines the balance between private and public space within walking distance of the home. Quarter-acre block housing with its large space standards (including large gardens), such as is common in Australia for example, places public and commercial facilities largely out of pedestrian access and effectively enforces car-dependency, particularly in contrast to apartment block or small-plot terraced housing. Increasing private space at the expense of public space also constitutes a strong supporting factor for more home-oriented living in general and domestic female labour roles in particular. It is probably safe to say that the single most important factor in determining the socio-spatial characteristics of a locality is the dwelling type that is dominant in it. This places housing at the centre of urban and regional research, rather than being, as so often is the case, a side issue of urban sociology.

But housing is not just embedded in locality as a socio-spatial structure. It is also embedded in the institutional structure of a society: the mesh of organisational and other institutional arrangements that have been evolved over extended periods of time to handle the financing, building, allocation, administration and maintenance of dwellings in the profuse variety of forms of provision that have been developed in any given society. These, of course, include the state in the multifarious forms in which it is involved in housing both directly and indirectly. But in the wider perspective of residence, housing is just one aspect. Residence is embedded in social structure in much more complex and strategic ways, profoundly influencing the social organisation of localities, and strongly affecting planning by both state agencies and other interests.

In all of this, the state looms large in a wide variety of forms and contexts, ranging from household social security provision, care of the aged, local planning, the provision of educational, health, and other public institutions, and, of course taxation of, and subsidies to, both households and dwellings. My contention is that the state must be treated as one of the agents in the embedding of housing and residence in social structure, with a degree of autonomy and its own interests as argued from the statist perspective. That is, representatives of the state do have broader concerns of social order and security that go beyond the sponsoring of sectional social interests. At the same time, influence over state apparatuses gives vested interests considerable power to define or redefine the social relations of residence, and the manner in which they are organised.

There is an absence of research which takes a broader residence-based perspective on housing. The closest approximation to this is the debate on consumption cleavages, and especially the work of Saunders (1978, 1990). Building on the housing classes debates, Saunders argues that owner occupied housing is a major source of social inequality because of its potential as a means of accumulating wealth. Saunders argues that the possession or otherwise of owner occupied housing comprises a major dividing line between those with and those without the ability to benefit from the social organisation of modern industrial society. Those without such housing become increasingly marginalised and dependent on the provision by the state of services of declining quality. Saunders' perspective is, then, one in which housing plays a central part and is embedded in social structure in crucial ways in relation to social class, the market and the welfare state. Despite the specific ideological slant of the analysis, Saunders' work therefore constitutes probably the most sociological and embedded treatment of housing at present.

But Saunders has an essentially intuitive understanding of the importance of housing in social structure, and does not explicitly theorise this. More specifically, he focuses on one form of tenure and on only one aspect of it; owner occupation as a source of wealth accumulation. He therefore does not fully appreciate the embeddedness of housing in its broadest sense and the far reaching ramifications of housing for social structure – socially, economically, and politically. In particular, he neglects the socio-spatial nature of housing and its importance for the organisation of social life and basic institutions such as the family. It is also unfortunate that Saunders does not locate his analysis in a broader comparative and historical explanatory framework about consumption cleavages in industrial societies.

RESIDENCE AND THE DYNAMICS OF SOCIAL CHANGE

This brings us back to the prime moving factors that are at work in structuring the social relations of residence. We have noted that in social structures there emerge characteristic forms of social organisation for the provision of housing and its wider residential implications. The question I wish to address in this context is what are the dynamics that underlie the specific configurations of social relations of residence of a particular society?

These social arrangements take an almost bewilderingly wide variety of forms. Take the financing of housing as one example. Private building companies who need construction finance, the state or local authority or other builders, including self-builders, have all different needs and requirements. In addition, financing requirements suited to owner occupiers buying a dwelling from a builder or from a prior occupier, or landlord are needed. All these and other needs are met by the development over long periods of time by the evolving of various institutional arrangements for the provision of finance. Many of these have modest beginnings. For example in lending by solicitors or estate agents, in temporary building societies and other co-operative forms of mutual aid, and in small finance companies or co-operatives. These develop and change over the generations until they form the modern system of finance of a country. Each country has its own characteristic array of systems of finance, and they are constantly changing today, partly in response to changes in other social relations of residence. What is it that decides the forms which ultimately become dominant in any given society?

Conflict between different vested interests must be central to such a process. But this on its own tells us little. It tells us nothing about how it comes about and what reasons there may be for some interests to succeed in becoming dominant over others. It is the *process whereby such domination is accomplished* that is of interest here. It is here that a neo-Gramscian approach combined with a negotiated order perspective has powerful explanatory force.

The institutional order of social structure is comprised of innumerable social arrangements which are made in the process of interpersonal interaction, as people struggle to create new forms of co-operation and to adjust to the new forms of co-operation that others have set up, thereby creating unintended consequences of those new forms of co-operation. But the ideas that form the basis of social action do not emerge from a vacuum. They are influenced by the social arrangements that were created out of past inter-action and can be influenced by the forms of co-operation that may be set up within existing social arrangements. Crucial in that process is the way in which ideas are channelled into certain directions and away from others by basic assumptions and agendas that have emotive appeal.

Take, for example, the act of setting up a new business. Most people in capitalist industrial societies would immediately think in terms of setting up a company and perhaps hiring employees. They may do so individually or in partnership with friends or relatives. But the idea of setting up a co-operative would probably not occur to many, and, for the few who do, the institutional arrangements are relatively poorly developed. Thus, the line of least resistance and the one with the most precedents and institutional advice, help and support would be the privatist way. That this is so is due to the existence of a privatist ideological dominance. There is nothing inherently inferior about co-operation as a form of social organisation for enterprise. The same principle holds in housing. House buying is organised round complex institutional arrangements designed to cater for individuals or couples who are buying. It is much harder for a group of people to buy a house, and even harder for such a group to have collective housing built. A fundamental way of changing social relations must therefore be to establish a counter-ideology, if possible as part of hegemony and backed up by political dominance, so that laws may be adapted and subsidies redirected in order to support the new forms. Emotional commitment constitutes a vital ingredient in this, most dramatically evidenced in political ideologies and manifested in public policy.

The institutional order of social structure is therefore to a large extent determined by the dominant ideology in civil society. Societies have dif-ferent institutional orders because over time different ideologies have become established as dominant, and so different social arrangements have been evolved. It is my contention that the privatism/collectivism distinction constitutes one of the major ideological cleavages in industrial societies and explains the process of divergence that has taken place between them. Residence plays a central role in this process, because of its huge impact on life styles and hence degrees of privatism or collectivism in social structures.

The divergence thesis is, however, a subsidiary theme in this book. It is posited partly to develop further my earlier work in this area but, more significantly, to illustrate the centrality of residence to social structure, and the importance of theorising housing in terms of residence by applying a conceptual framework to it (in this case combining neo-Gramscian and negotiated order theory). I want now to return to the central theme, namely, developing a sociology of residence which transcends housing studies, as narrowly conceived, and is adequately embedded in social structure.

TOWARD A SOCIOLOGY OF RESIDENCE

I have proposed that residence can be conceived in terms of a series of chinese boxes. At the centre is the household, which is within the dwelling, which is in a locality (see Figure 5.1). These in turn are nested in an institutional structure. But this is an overly spatial perspective which tends to play down the social dimension of residence (in particular the social institutions that are centrally involved in issues of residence but are physically located outside of the locality). Another way of conceiving this relationship is in terms of the relationship between the socio-spatial dimension on the one hand and base unit to collective unit levels of analysis on the other. This can be schematically represented as in Figure 9.1.

Figure 9.1 The components of residence

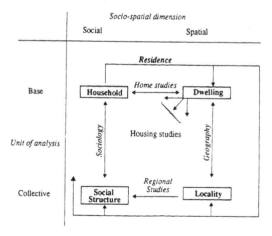

Base and collective are not the same as micro–macro, but rather distinguish between the basic units of analysis (individual and dwelling) and all other higher levels (which may range in social terms from small groups through organisations to state-bounded societies and beyond, or in spatial terms from block, neighbourhood, suburb, city, region, etc.). The vertical relationships represent the social and spatial disciplines: the left hand column representing the social dimension, as studied in sociology, from the individual to social structure, and the right hand column represents the spatial dimension as studied in geography, ranging from dwelling to locality. The horizontal relationships represent other social science approaches linking the social and spatial dimensions at both base unit and collective unit levels.

Using the unproblematised definition of the substantive focus of housing studies in terms of 'shelter' (see Chapter 1), housing studies takes dwellings as its starting point and attempts to relate dwellings to both individual and collective levels of social analysis, and to collective spatial units. In practice, the extent to which the other three dimension elements constitute a subject for housing research varies enormously. Some housing research is heavily locality based. Some is oriented primarily towards abstract, spatially non-specific institutional analysis (social structure). Yet other studies are oriented towards the individual or household. Some research is oriented towards two dimensions in interaction, for example, residential mobility studies are often oriented towards the individual and locality dimensions ('who moves where'). But in general, all this is left unexplicated. It is this that makes housing studies such a mish-mash of different focuses and interests.

Let us now consider how residence can be conceptualised in terms of Figure 9.1. Residence embraces all four dimensions in an interactive relationship that cannot be adequately captured in two dimensions on paper. But relating Figure 9.1 to Figure 5.1, the three chinese boxes in Figure 5.1 represent the three first dimensions in Figure 9.1 following the arrow from individual onwards. In the most general terms, analysis moves from the social to the spatial dimension and from the individual to the collective level (or vice versa), and back again. Analysis therefore consists of moments that are predominantly social and moments that are predominantly spatial, each with base focuses and collective focuses. This is an oversimplified representation of the research process, of course, but captures the main dimensions of the study of residence and constitutes at least a starting point for developing a sociology of residence.

CONCLUSIONS

In the Introduction to this book I traced, as many have done before me, the emergence of an interest in theorising housing back to Rex and Moore (1967) and the concept of housing classes. Since then, housing has continued to loom large in explanations of social inequality, but it has remained a sort of 'factor X', a vital external ingredient that has to be added to a sociological analysis to help to understand class and inequality issues but that always seems to elude theorising.

Yet the concept of housing classes, important though it has been for rising theoretical awareness among housing researchers, has resulted in something of a wild goose chase, never yielding the promise that it seemed to hold out. I believe that this is because of a fundamental misreading of what the originators of this concept were trying to do.

Rex and Moore's main interest was not in some abstract conceptualisation of housing classes *per se*, but in using it as part of their attempt to develop a wider sociology of the city based on socio-spatial dimensions *in which housing figured as the central and organising factor*. The concluding chapter of their book, which is titled 'the sociology of the zone of transition' and where the concept of housing classes is developed, places class conflict over access to housing as a central feature of cities. Modifying the tradition of the Chicago School of urban sociology, they identify different parts of cities in which different forms of housing predominate, and the zone of transition comprises the focus for ethnic conflict over housing.

The housing classes debate that followed ranged over many of the side issues, such as the taxonomy of housing classes, the existence or otherwise of a common set of values, and the significance of tenure. The most promising line of development was that of Pahl's (1969) concept of urban managerialism, in terms of the main actors in the allocation of housing and other key resources. But even this was only a partial development of the original thesis and the principal point about housing being fundamental to the social organisation of urban areas was missed.

This misreading of the significance of housing in Rex and Moore's analysis is in some part the fault of the way in which the concept was presented. Theoretically, the final chapter of *Race, Community and Conflict* was very under-developed and somewhat confused, starting out to present a theoretical model of urban processes as they bear on twilight zones of transition and ending by appealing to planners and social workers to deal with its problems. In the middle of this the concept of housing classes stands out as the principal substantive contribution of the chapter.

There are strong parallels here with the way in which my earlier work was received. Commentators seized on the housing issues and ignored the wider social structural ones in which housing was embedded. I do not believe that this is a coincidence. The conceptual integration of housing into social structure contains a perceptual trap that is very easy to fall into. Housing is such an 'obvious' and unproblematic phenomenon. This also means, unfortunately, that dwellings as objects can be easily pushed aside for analytical purposes and treated as just another material possession. This simple common sense – and undersocialised – understanding of housing therefore conceals the enormous structural implications of residence: implications which should place residence at the heart of social theory.

10 Conclusions

In these concluding words I want to look beyond this book and to consider possible ways forward for a theorised housing studies. If I am to follow my own suggestions concerning the need for greater reflexivity, the starting point for such an exercise must be a critique of my own argument. The maxim that 'books are never completed – they are abandoned' is particularly apt in this case. Theorising is a process of interaction, with oneself and with other people in the form of both an internal dialogue and a dialogue between one's own written words and reflections on them. A book of this sort therefore represents a tiny negotiated order frozen in a moment in time. Even while the last changes are being made, the dialogue goes on and the grounds of the debate change as the 'final' text constitutes the starting point for further development in which new perspectives are introduced and old ones toned down or abandoned in the course of both interpersonal and 'text-mediated' interaction.

There are many problems in writing a book such as this, particularly as a result of the breadth of the topic. In the first place it encourages a degree of dilettantism. Nothing can be dealt with in any depth and there is a temptation to give a gloss to, or even shy away from addressing, really thorny issues. This is particularly evident in my handling of the divergence thesis. An adequate treatment of this would have required at the very least a far more detailed definition of the collective/private distinction, followed by analysis of time-series statistical data on international patterns of development in terms of the collective/private distinction, and preferably also at least two detailed national case studies. A further problem is the fact that the book is fairly loosely integrated. It straddles a wide range of issues and lacks the tightly organised character of an empirical study comprising theory, methods, findings and conclusions, or a theoretical treatise exhaustively pursuing a single idea to its conclusion. The unifying theme is there, but more as an undercurrent manifested in a diverse range of topics. Finally, the book can be criticised for its hermeneutic bias which can be seen as divisive, particularly since housing research is dominated by the positivist tradition.

My response to such criticisms would be that the form of the discussion is largely deliberate. I have tried to open up a wide range of issues and to treat them provocatively in order to stimulate debate. And while I do base much – though by no means all – of my own work on the hermeneutic tradition of sociology, I have employed it here not to proselytise for its being used as the basis of theorising housing but catalytically, as a means of provoking reflexivity, since this aspect is grossly neglected in much of the literature. The use of an incomplete case study in Part 3 emphasises this and provides a clue as to where to go from here. Clearly the next task is to outline a possible programme of research that could grow naturally out of my arguments. This can be divided into two main kinds of studies: epistemological and empirical.

Work which develops the epistemology of housing is needed in several respects. The most important is to develop a satisfactory social structural focus for housing research. I believe that some concept along the lines of 'residence' provides a useful starting point. Such a subject focus would also provide a conceptual link between housing studies and the other social sciences. It is likely that such a concept could constitute a key element in socio-spatial studies. But this is a huge task that will take many contributions and much debate to come to some satisfactory resolution. A good parallel is the way in which the concept of welfare has moved from an unreflexive focus on statutory provision to a complex view of welfare. But this has taken twenty years of debate, many books, and a number of false starts. If similar progress is to be made in housing studies, perhaps based on some concept of residence, an equally long gestation period is quite possible. Nor do I rule out the possibility that this start may prove false and that better ways forward might be devised. But at least it is a start.

A critical approach to the social construction of housing research is another important area that has been neglected and needs work. The example in Chapter 8 of the manner in which ideological assumptions are built into housing statistics to justify building either houses or flats shows this clearly. We have been far too uncritical of the way in which housing statistics have been manufactured, and have taken an overly positivistic and unproblematised attitude towards them.

Work also needs to be done on the social construction of housing problems. Wright Mills distinguished between 'the personal troubles of milieu' and 'the public issues of social structure', underlining that the nature of the latter is often debatable (Mills, 1959: 14–15). One reason why housing researchers have been so heavily policy-oriented is that there has tended to be an unproblematic acceptance of the definitions of policy-makers as to what does or does not constitute a housing problem. This is not to say that researchers should ignore policy-makers' definitions or constantly attempt to de-legitimise or undermine them. Rather, it is to come to a researcher-determined definition of a housing problem.

These sorts of studies which question underlying taken-for-granted assumptions within housing research are well established in other areas of the social sciences. They derive from an attitude of profound scepticism which is grounded in a critical and radical social science tradition, drawing on both Marxist and non-Marxist traditions, and which has yet to find resonance in housing research. The choice need not be between Marxist heavy duty structuralism and positivistic – and sometimes ultra-conservative – Weberianism, or between a provision and consumption approach. Indeed, the choice is increasingly much more varied than this and is widening continually. But a genuine alternative to these, which is a viable perspective in its own right and in which critical reflexive analysis figures centrally, is not only possible but also, I argue, highly fruitful. The emergence of such a radical housing studies would be a healthy sign that housing research is beginning to ground itelf in reflexivity and self-critique, which I would argue is an essential prerequisite to the development of a theoretically strong housing studies.

A third major area for development must be empirical studies in housing which are grounded in existing conceptual debates in the social sciences. Such research is increasingly being carried out, but much housing research is still insular in outlook. In its extreme form, the approach I am arguing for would result in the abolition of housing studies and its reintegration into social science in which the dimension of residence comprises a major conceptual focus of social science research. This caricature has a grain of truth in it but is greatly overdrawn. As with all else, the issue resolves itself into one of balance. And as with all questions of balance there is always considerable variation between researchers as to what the 'right' balance should be. My argument therefore resolves itself into a plea for more integration of empirical housing research into mainstream social science. Housing research can and should both benefit more from debates in the social sciences and – equally important – contribute to them by testing propositions, for example about convergence/divergence, the welfare state, labour movement theory, the autonomy of the state and corporatism.

Over and above this there is a real need for housing researchers to argue that the social sciences should take housing issues more seriously than they do in their explanations and theories. This will happen partly as a result of housing researchers testing general propositions with housing data. But it also requires housing researchers to engage actively in current debates which, on the surface of it, would appear to have little directly to do with housing. It is here that an embedded and conceptualised subject focus for housing studies can play a decisive part in reintegrating housing research into social theory and thence back into broader social issues. This brings the discussion full circle to the first, epistemological task that is needed beyond this book concerning the concept of residence, or some similar conceptualised way of conceiving housing. I would argue that such a task provides the main challenge to housing researchers in the immediate future.

Notes

4 A CRITIQUE OF UNILINEALISM IN COMPARATIVE HOUSING RESEARCH

1 It is interesting to note that Kerr later appeared to move closer to the convergence interpretation of his and his colleagues thesis (Kerr, 1983).

5 HOUSING AND COMPARATIVE WELFARE RESEARCH

1 It is testimony to the power of ideology that this is not how the tenures are commonly perceived.

6 IDEOLOGY AND DIVERGENT SOCIAL STRUCTURES

1 Sometimes Gramsci uses the term bloc to refer to the integration of levels (structure and superstructure; Gramsci, 1971: 137), elsewhere in the abstract sense of integrating history and philosophy (Gramsci, 1971: 345).

2 Emotion has been a grossly neglected factor in social science research until recently. The reason for this lies mainly in the theories of industrialisation that remain to this day dominant in sociology, in which traditional society is held to be based on primary affective relationships while modern society is based on instrumental secondary relationships (Hammond, 1983), but also partly on the extreme patriarchal structure of academia, reflecting masculine values that deny the validity of emotions (Jagger, 1989). Collins, and others, argue convincingly that emotional attachment is a key ingredient in making all forms of social relationships viable.

3 This distinction may be compared to the first two of a threefold distinction between 'paradigm', 'metaphor', and 'puzzle solving' (Morgan, 1980). 'Paradigm' (such as functionalism) refers to alternative world views, while 'metaphor' refers to the variety of images (such as mechanistic, organistic, cybernetic) associated with the paradigm.

7 DIVERGENT SOCIAL STRUCTURES AND RESIDENCE

1   Schmidt (1989a) also notes the exception of Switzerland in some of his data.

8 THE POLITICAL CONSTRUCTION OF COLLECTIVE RESIDENCE: THE CASE OF SWEDEN

1   Housing is a particularly good example of hegemonic corporatism at work, particularly in the inclusion of private landlords in the subsidy system.
2   The expression "society-building" figures in both polemics for the People's Home (SOU, 1945) and analysis of Swedish People's Home ideology (Carlestam, 1986).
3   This part of the chapter is based on Franzén and Sandstedt (1981) which provides the best account of ideology and urban planning in Sweden.
4   It might be noted here that Segerstedt was not a social democrat. This is indicative of the wide appeal of the dominant ideology, in its ability to attract intellectual support from a wide range of interests.
5   In the light of this argument it is ironic that the success of the policy of supporting the commodification of female labour by high levels of collective provision of facilities, notably childcare, did not solve the problem of double labour but merely perpetuated it.
6   Major changes have taken place in government policy towards urban renewal, however, with an ideologically motivated shift towards privatist solutions (see Parkinson, 1989).

References

Aaronovitch, S. (1961) *The Ruling Class: a study of British finance capital*, Lawrence & Wishart, London.

Abercrombie, N., Hill, S., and Turner, B.S. (1980) *The Dominant Ideology Thesis*, Allen & Unwin, London.

Abercrombie, N., Hill, S., and Turner, B.S. (1986) *Sovereign Individuals of Capitalism*, Allen & Unwin, London.

Abrams, Charles (1964) *Man's struggle for shelter in an urbanizing world*, MIT Press, Cambridge, MA.

Almond, Gabriel (1988) 'The Return to the State', *American Political Science Review* (September) 82, 3: 875–901 (or 853–74).

Atkinson, J.M. (1978) *Discovering Suicide: studies in the social organization of sudden death*, Macmillan, London.

Bachrach, Peter and Baratz, Peter (1962) 'The two faces of power', *American Political Science Review* (December) 56, 4: 947–52.

Bachrach, Peter and Baratz, Peter (1963) 'Decisions and non-decisions: an analytical framework', American Political Science Review (September) 57, 3: 632–42.

Baggueley, P., Lawson, J., Shapiro, D., Urry, J., Walby, S. and Ward, A. (1990) *Restructuring: people, class and gender*, Sage, London.

Ball, Michael (1983) *Housing Policy and Economic Power: the political economy of owner occupation*, Macmillan, London.

Ball, M., Harloe, M., and Martens, M. (1988) *Housing and Social Change in Europe and the USA*, Routledge, London.

Ball, Richard A. and Lilly, Robert (1982) 'The Menace of Margarine: the rise and fall of a social problem', *Social Problems* (June) 29, 5: 488–90.

Banfield, Edward C. (1958) *The moral basis of a backward society*, Free Press, New York.

Barrington Moore jun. (1966) *The Social Origins of Dictatorship and Democracy: lord and peasant in the making of the modern world*, Beacon Press, Boston, MA.

Becker, Howard (1967) 'Whose side are we on?', *Social Problems* (Winter) 14: 239–47.

Bell, Colin (1977) 'On housing classes', *Australian and New Zealand Journal of Sociology* (February) 3, 1: 36–40.

Bendix, Reinhart (1964) *Nationbuilding and Citizenship*, Wiley, New York.

Berger, Peter and Kellner, Hansfried (1970) 'Marriage and the construction of reality: an exercise in the microsociology of knowledge' in Dreitzel, Hans (ed.) *Recent Sociology* No. 2, Macmillan, London, 50–72.

Berger, Peter and Luckmann, Thomas (1966) *The Social Construction of Reality: a treatise in the sociology of knowledge*, Penguin, Harmondsworth, Middlesex.

Bhaskar, Roy (1975) *A Realist Theory of Social Science*, Alma, Leeds.

Black, Donald J. (1970) 'Production of crime rates', *American Sociological Review* (August) 35: 733–48.

Block, Fred (1987) *Revising State Theory*, Temple University Press, Philadelphia, PA.

Bloomfield, J. (ed.) (1977) *Class, Hegemony, Party*, Lawrence & Wishart, London.

Blumer, Herbert (1969) *Symbolic Interactionism: perspective and method*, Prentice Hall, Englewood Cliffs, NJ.

Boudon, Raymond (1986) *Theories of social change: a critical appraisal*, Polity Press, Cambridge (trans. J.C. Whitehouse).

Carlestam, Gösta (1986) *Society Builders by the Great Lake: a place in the outback becomes a town in the world. A discussion of the steeltown of Sandviken – and on goals and means in the social planning of Sweden 1862–1984* (in Swedish), National Swedish Institute for Building Research, Gävle.

Carlstein, T. (1981) 'The sociology of structuration in time and space: a time-geographical assessment of Giddens' theory', *Swedish Geographical Yearbook* 57: 41–57.

Carrier, John and Kendall, Ian (1977) 'Social Administration as Social Science', pp. 25–32 in Helmuth Heisler (ed.) *Foundations of Social Administration*, Macmillan, London.

Castells, Manuel (1977) *The Urban Question: a Marxist approach*, Matthew Arnold, London.

Castles, Francis G. (1978) *The Social Democratic Image of Society*, Routledge, London.

Castles, Francis G. (1982) 'The impact of parties on public expenditure', pp. 21–96 in Francis G. Castles (ed.) *The Impact of Parties: politics and policies in democratic capitalist states*, Sage, London.

Castles, Francis G. (1985) *The Working Class and Welfare: reflections on the political development of the welfare state in Australia and New Zealand, 1890–1980*, Allen & Unwin, Wellington, NZ.

Cicourel, Aaron V. (1964) *Method and Measurement in Sociology*, Free Press, New York.

Clark, Gordon L. and Dear, Michael (1984) *State Apparatus: structures and language of legitimacy*, Allen & Unwin, Boston, MA.

Clarke, Simon and Ginsburg, Norman (1975) 'The Political Economy of Housing', *Political Economy and the Housing Question*, Political Economy of Housing Workshop, Conference of Socialist Economists, London.

Clarke, Simon and Ginsburg, Norman (1976) 'The political economy of housing', *Kapitalistate* (Summer) No. 4/5 pp. 66–99.

Cohen, Stanley (1970) *Folk Devils and Moral Panics*, Paladin, London.

Cohen, Stanley and Young, Jock (eds) (1973) *The Manufacture of News: social problems, deviance and the mass media*, Constable, London.

Collins, Randall (1975) *Conflict Sociology: toward an explanatory science*, Academic Press, New York.

Collins, Randall (1981) 'The microfoundations of macrosociology', *American Journal of Sociology* (March) 86, 5: 984–1014.

Collins, Randall (1984) 'The role of emotion in social structure', pp. 385–96 in

174 References

Klaus R. Scherer and Paul Ekman (eds) *Approaches to Emotion*, Lawrence Erlbaum Associates, Hillsdale, NJ.

Collins, Randall (1986) *Weberian Sociological Theory*, Cambridge University Press, Cambridge.

Collins, Randall (1987a) 'A micro–macro theory of intellectual creativity: the case of German idealist philosophy', *Sociological Theory* (Spring) 5: 47–69.

Collins, Randall (1987b) 'Interaction ritual chains: the micro–macro connection as an empirically based theoretical problem', pp. 192–205 in Jeffrey C. Alexander, Bernhard Geisen, Richar Münch, and Neil J. Smelser (eds) *The Micro–Macro Link*, University of California Press, Berkeley, CA.

Collins, Randall (1989) 'Toward a neo-Meadian sociology of mind', *Symbolic Interaction* (Spring) 12, 1: 1–32.

Couper, Mary and Brindley, Timothy (1975) 'Housing classes and housing values', *Sociological Review* (August) 23, 3: 563–76.

Crenson, Maurice (1971) *The Unpolitics of Air Pollution: a study of non-decision making in the cities*, John Hopkins University Press, Baltimore, MD.

Cullingworth, J.B. (1960) *Housing needs and planning policy: a restatement of the problem of housing need and 'overspill' in England and Wales*, Routledge & Kegan Paul, London.

Dahl, Robert A. (1962) *Who Governs?*, Yale University Press, New Haven, CT.

Davidson, Alexander (1989) *Two models of welfare: the origins and development of the welfare state in Sweden and New Zealand 1888–1988*, Political Science Association (publication 108), Uppsala University.

Day, Robert and Day, Jo Anne V. (1977) 'A review of the current state of negotiated order theory: an appreciation and critique', *Sociological Quarterly* (Winter) 18, 1: 126–42.

de Neufville, J.I. and Barton, S.E. (1987) 'Myths and the definition of policy problems', *Policy Sciences* 20, 181–201.

Denzin, Norman K. (1970) *The Research Act: a theoretical introduction to sociological methods*, McGraw-Hill, New York.

Dickens, Peter (1989) 'Human Nature, Society and the Home', *Housing Studies* (October) 4, 4: 227–37.

Dickens, P., Duncan, S., Goodwin, M., and Gray, F. (1985) *State, Housing and Localities*, Metheun, London.

Ditton, Jason (1979) *Contrology: beyond the new criminology*, Macmillan, London.

Donnison, D.V. (1967) *The Government of Housing*, Penguin, Harmondsworth, Middlesex.

Donnison, David and Ungerson, Clare (1982) *Housing Policy*, Penguin, Harmondsworth, Middlesex.

Douglas, Jack D. (1967) *The Social Meaning of Suicide*, Princeton University Press, Princeton, NJ.

Douglas, Mary and Wildavsky, Aaron (1982) *Risk and Culture*, University of California Press, Berkeley, CA.

Duncan, S.S. (1981) 'Housing policy, the methodology of levels, and urban research: the case of Castells', *International Journal of Urban and Regional Research* (June) 5, 2: 231–52.

Duncan, Simon and Goodwin, Mark (1988) *The Local State and Uneven Development*, Polity Press, London.

Duncan, Simon and Savage, Mike (1989) 'Space, scale and locality', *Antipode* (December) 21, 3: 179–206.

Dunleavy, Patrick (1980) *Urban Political Analysis: the politics of collective consumption*, Macmillan, London.

Dunleavy, Patrick (1981) *The Politics of Mass Housing in Britain 1945–1975: a study of corporate power, and professional influence in the welfare state*, Clarendon Press, Oxford.

Dunning, E.G. and Hopper, E.I. (1966) 'Industrialisation and the problem of convergence: a critical note', *Sociological Review* (July) 14, 1: 163–86.

Durkheim, Emile (1952) *Suicide*, Routledge, London.

Edelman, Murray (1971) *Politics as symbolic action: mass arousal and acquiescence*, Markham, Chicago, IL.

Edelman, Murray (1977) *Political Language: words that succeed and policies that fail*, Academic Press, New York.

Elzinga, A., Geiger, R.L., and Wittrock, B. (1985) 'Research bureaucracy and the drift of epistemic criteria', pp. 191–220 in B. Wittrock and A. Elzinga (eds) *The University Research System: the public policies of the home of science*, Almqvist & Wicksell, Stockholm.

Eriksson, Jan and Lindqvist, Margareta (1983) 'A new overcrowding norm?', National Swedish Institute for Building Research Working Paper.

Esping-Andersen, Gøsta (1985) *Politics against Markets: the social democratic road to power*, Princeton University Press, Princeton, NJ.

Esping-Andersen, Gøsta (1987) 'The comparison of policy regimes', pp. 3–12 in Rein *et al.* (1987).

Esping-Andersen, Gøsta (1990) *The Three Worlds of Welfare Capitalism*, Polity Press, Cambridge.

Esping-Andersen, G., Friedland, R., and Wright, E.O. (1976) 'Modes of class struggle and the capitalist state', *Kapitalistate* (Summer) 4/5, 186–220.

Evans, P.B., Rueschemeyer, D., and Skocpol, T. (eds) (1985) *Bringing the State Back in*, Cambridge University Press, Cambridge.

Fabbrini, Sergio (1988) 'The Return to the State: critique', *American Political Science Review* (September) 82, 3: 891–901.

Fine, Gary Alan (1984) 'Negotiated orders and organizational culture', *Annual Review of Sociology* 10: 239–64.

Fishman, Mark (1978) 'Crime Waves as Ideology', *Social Problems* (June) 25, 5: 531–43.

Forder, A., Caslin, T., Ponton, G., and Walklate, S. (1984) *Theories of Welfare*, Routledge, London.

Forrest, Ray and Lloyd, John (1978) 'Theories of the capitalist state: the implications for policy research', *Papers in Urban and Regional Studies* (June) 2: 28–45.

Forrest, Ray and Murie, Alan (1985) *An Unreasonable Act? Central–local government conflict and the Housing Act 1980*, SAUS Studies No. 1, School for Advanced Urban Studies, Bristol University.

Forrest, Ray and Murie, Alan (1988) 'The social division of housing subsidies', *Critical Social Policy* (Autumn) 8, 2: 83–93.

Frankenberg, Ronald (1966) *Communities in Britain: social life in town and country*, Penguin, Harmondsworth, Middlesex.

Franzén, Mats and Sandstedt, Eva (1981) *Neighbourhoods and town planning: state and building in postwar Sweden* (in Swedish), Sociology Studies No. 17, Uppsala University.

Garfinkel, Harold (1967) *Studies in Ethnomethodology*, Prentice Hall, Englewood Cliffs, NJ.

Garland, David (1990) 'Frameworks of inquiry in the sociology of punishment', *British Journal of Sociology* (March) 41, 1: 1–15.

George, Vic and Wilding, Paul (1976) *Ideology and Social Welfare*, Routledge, London.

Giddens, Anthony (1979) *Central Problems in Social Theory: action, structure and contradiction in social theory*, London, Macmillan.

Giddens, Anthony (1982) *Sociology: a brief but critical introduction*, Harbrace, London.

Giddens, Anthony (1985) *The Nation State and Violence*, Polity Press, Cambridge.

Glaser, Barney and Strauss, Anselm L. (1967) *The Discovery of Grounded Theory*, Aldine, Chicago, IL.

Goldthorpe, John H. (1964) 'Social stratification in industrial society' in Paul Halmos (ed.) *The Development of Industrial Societies*, Sociological Review Monograph No. 8.

Gough, Ian (1979) *The Political Economy of the Welfare State*, Macmillan, London.

Gramsci, Antonio (1971) *Selections from the Prison Notebooks* (eds Q. Hoare and G. Nowell-Smith), Lawrence & Wishart, London.

Granovetter, Mark (1985) 'Economic action and social structure: the problem of embeddedness', *American Journal of Sociology* (November) 91, 3: 481–510.

Green, Arnold W. (1975) *Social Problems: arena of conflict*, McGraw-Hill, New York.

Gregory, Derek and Urry, John (1985) *Social Relations and Spatial Structures*, Macmillan, London.

Gurney, Craig (1990) 'The meaning of home in the decade of owner occupation: towards an experiential research agenda' (Working Paper 88), School for Advanced Urban Studies, Bristol University.

Gusfield, Joseph (1963) *Symbolic Crusade: status politics and the American temperance movement*, University of Illinois Press, Urbana, IL.

Hägerstrand, T. (1970) 'What about people in regional science?', *Papers of the Regional Science Association* 24: 1–21.

Hägerstrand, T. (1974) 'Time-geographic description: purpose and postulates', *Swedish Geographical Yearbook* 50: 86–94.

Hall, Peter M. and Spencer-Hall, Dee Ann (1982) 'The social conditions of the negotiated order', *Urban Life* (October) 11, 3: 328–49.

Hammond, Michael (1983) 'The sociology of emotions and the history of social differentiation', pp. 90–129 in Randall Collins (ed.) *Sociological Theory 1983*, Jossey Bass, San Francisco, CA.

Harloe, Michael (ed.) (1975) *Proceedings of the Conference on Urban Change and Conflict*, Centre for Environmental Studies, London (cyclostyled).

Harloe, Michael (ed.) (1977) *Captive Cities: studies in the political economy of cities and regions*, Wiley, New York.

Harloe, Michael and Martens, Maartje (1984) 'Comparative housing research', *Journal of Social Policy* 13, 3: 255–77.

Harrison, M.L. (1986) 'Consumption and urban theory: an alternative approach based on the social division of welfare', *International Journal of Urban and Regional Research* (June) 10, 2: 232–42.

Hayden, Dolores (1981) *The Great Domestic Revolution: a history of feminist designs for American homes, neighborhoods and cities*, MIT Press, Cambridge, MA.

Hayward, David (1986) 'The Great Australian Dream reconsidered: a review of Kemeny', *Housing Studies* (October) 1, 4: 210–19.

Heclo, Hugh (1974) *Modern social politics in Britain and Sweden: from relief to income maintenance*, Yale University Press, New Haven, CT.

Heclo, Hugh and Madsen, Henrik (1987) *Policy and Politics in Sweden: principled pragmatism*, Temple University Press, Philadelphia, PA.

Heclo, Hugh and Wildavsky; Aaron (1981) *The Private Government of Public Money: community and policy inside British politics*, Macmillan, London.

Held, David (1984) *Political theory and the modern state: essays on state, power and democracy*, Polity Press, Cambridge.

Hicks, A., Friedland, R., and Johnson, E. (1978) 'Class power and state policy', *American Sociological Review* 43, 2: 302–15.

Hilgartner, Stephen and Bosk, Charles L. (1988) 'The rise and fall of social problems: a public arenas model', *American Journal of Sociology* (July) 94, 1: 53–78.

Hillier, Bill and Hanson, Julienne (1984) *The social logic of space*, Cambridge University Press, Cambridge.

Hindess, Barry (1973) *The use of official studies in sociology: a critique of positivism and ethnomethodology*, Macmillan, London.

Hirdman, Yvonne (1989) *Putting life to rights: studies in Swedish People's Home Policy* (in Swedish), Carlssons, Stockholm.

Hunter, Floyd (1953) *Community Power Structure: a study of decision-makers*, University of North Carolina Press, Chapel Hill, NC.

Jagger, Alison M. (1989) 'Love and knowledge: emotion in feminist epistemology', *Inquiry* 1989, 32, 151–76.

Jessop, Bob (1982) *The Capitalist State: Marxist theories and methods*, Martin Robertson, Oxford.

Johnson, Norman (1987) *The Welfare State in Transition: the theory and practice of welfare pluralism*, Wheatsheaf, Brighton.

Kemeny, Jim (1976a) *An Interactionist Approach to Macrosociology*, Sociology Monograph No. 10, University of Gothenburg.

Kemeny, Jim (1976b) 'Perspectives on the micro–macro distinction', *Sociological Review* (November) 24, 4: 732–52.

Kemeny, Jim (1977) 'A political sociology of home ownership in Australia', *Australian and New Zealand Journal of Sociology* (February), 13, 1: 47–52.

Kemeny, Jim (1978) 'Forms of tenure and social structure: a comparison of owning and renting in Australia and Sweden', *British Journal of Sociology* (March) 29, 1: 41–56.

Kemeny, Jim (1980) 'Home Ownership and Privatisation', *International Journal of Urban and Regional Research* (September) 4, 3: 372–88.

Kemeny, Jim (1981) *The Myth of Home Ownership: public versus private choices in housing tenure*, Routledge, London.

Kemeny, Jim (1983a) *The Great Australian Nightmare: a critique of the home ownership ideology*, Georgian House, Melbourne.

Kemeny, Jim (1983b) 'Professional ideologies and organisational structure: tanks and the military', *European Journal of Sociology* 24: 223–40.

Kemeny, Jim (1984) 'The social construction of housing facts', *Scandinavian Housing and Planning Research* (August), 1, 3: 149–64.

Kemeny, Jim (1987a) 'Towards a theorised housing studies: a critique of the provision thesis', *Housing Studies* (October) 2, 4: 249–60.

Kemeny, Jim (1987b) 'The neglected individual in urban studies', *National Swedish Institute for Building Research Working Paper* Gävle.

Kemeny, Jim (1988) 'Defining housing reality: ideological hegemony and power in housing research', *Housing Studies* (October) 3, 4: 205–16.

Kerr, Clark (1983) *The future of industrial societies: convergence or continuing diversity*, Harvard University Press, Cambridge, MA.

Kerr, C., Dunlop, J.T., Harbison, F.H., and Myers, C.A. (1960) *Industrialism and Industrial Man: the problems of labor and management in economic growth*, Harvard University Press, Cambridge, MA.

Kitsuse, John I. and Cicourel, Aaron V. (1963) 'A note on the use of official statistics', *Social Problems* (Fall) 12, 2: 131–9.

Kitsuse, John I. and Schneider, Joseph (eds) (1982) *Advances in the Sociology of Social Problems*, Ablex, Norwood, NJ.

Kitsuse, John I. and Spector, Malcolm (1973) 'Toward a sociology of social problems: social conditions, value judgements and social problems', *Social Problems* (Spring) 20, 4: 407–19.

Knorr-Cetina, K. and Cicourel, A.V. (eds) (1981) *Advances in Social Theory and Methodology: toward an integration of macro- and micro-sociologies*, Routledge & Kegan Paul, Boston, MA.

Korpi, Walter (1978) *The Working Class in Welfare Capitalism*, Routledge & Kegan Paul, London.

Korpi, Walter (1983) *The Democratic Class Struggle*, Routledge & Kegan Paul, London.

Laclau, E. (1977) *Politics and Ideology in Marxist Theory: capitalism, fascism, populism*, New Left Books, London.

Laclau, Ernesto and Mouffe, Chantal (1985) *Hegemony and Socialist Strategy*, New Left Books, London.

Lambert, J., Paris, C., and Blackaby, B. (1978) *Housing Policy and the State: allocation access and control*, Macmillan, London.

Larrain, Jorge (1979) *The concept of ideology*, Hutchinson, London.

Larrain, Jorge (1983) *Marxism and Ideology*, Humanities Press, Highlands, NJ.

Latour, Bruno (1987) *Science in action: how to follow scientists and engineers through society*, Open University Press, Milton Keynes.

Latour, Bruno, Woolgar, Steven (1979) *Laboratory Life: the social construction of scientific facts*, Sage, Beverley Hills, CA.

Lenntorp, S. (1976) 'Paths in space–time environments: a time-geographic study of movement possibilities of individuals', *Lund Studies in Geography* Series 3 No. 44, University of Lund.

Lipset, Seymour Martin (1959) *Agrarian Socialism: the Coöperative Common-wealth Federation in Saskatchewan: a study in political sociology*, University of California Press, Berkeley, CA.

Lowi, Theodore J. (1988) 'The Return to the State: critique' *American Political Science Review* (September) 82, 3: 885–91.

Lundqvist, Lennart J. (1988) *Housing policy and tenures in Sweden: the quest for neutrality*, Gower, Aldershot.

Lundqvist, Lennart J. (1989a) 'Privatisation: towards a concept for comparative policy analysis', *Journal of Public Policy* 8, 1: 1–19.

Lundqvist, Lennart J. (1989b) 'Explaining privatisation: notes towards a predictive theory', *Scandinavian Political Studies* 12, 2: 129–45.

Maines, David R. (1978) 'Structural parameters and negotiated orders: comment on Benson and Day and Day', *Sociological Quarterly* (Summer) 19, 3: 491–6.

Maines, David R. (1982) 'In search of mesostructure: studies in the negotiated order', *Urban Life* (October) 11, 3: 267–79.

Mankoff, Milton (ed.) (1972) *The Poverty of Progress: the Political Economy of American Social Problems*, New York, Holt, Rhinehart and Winston.

Mann, Michael (1984) 'The autonomous power of the state: its origins, mechanisms and reality', *European Journal of Sociology* 25, 185–213.

Mann, Michael (1987) 'Ruling class strategies and citizenship', *Sociology* (August) 21, 3: 339–54.

Marshall, T.H. (1950) *Citizenship and Social Class*, Cambridge University Press, Cambridge.

Martin, Wilfred B.W. (1976) *The Negotiated Order of the School*, Maclean-Hunter Press, London.

Maruo, Naomi (1986) 'The development of the welfare mix in Japan', pp. 64–79 in Rose and Shiratori (1986).

Mauss, Armand L. (1975) *Social Problems as Social Movements*, Lippincott, Philadelphia, PA.

Mead, George Herbert (1934) *Mind, Self and Society: from the standpoint of a social behaviorist*, Chicago University Press, Chicago, IL.

Milibrand, Ralph (1969) *The State in Capitalist Society*, Weidenfeld & Nicholson, London.

Mills, C. Wright (1956) *The Power Elite*, Oxford University Press, New York.

Mills, C. Wright (1959) *The sociological imagination*, Oxford University Press, New York (Pelican 1970).

Mishra, R. (1981) *Society and Social Policy: theories and practice of welfare*, Macmillan, London.

Mollenkopf, J. and Pynoos, J. (1972) 'Property, politics and local housing policy', *Politics and Society* (Summer) 2, 2: 407–32.

Morgan, Gareth (1980)'Paradigms, metaphors, and puzzle solving in organization theory', *Administrative Science Quarterly* (December) 25, 4: 605–22.

Nordlinger, Eric (1981) *On the Autonomy of the Democratic State*, Harvard University Press, Cambridge, MA.

Nordlinger, Eric (1987) 'Taking the State Seriously', in Myron Weiner and Samuel Huntington (eds) *Understanding Political Development*, Little, Brown, Boston, MA.

Nordlinger, Eric (1988) 'The Return to the State: critique', *American Political Science Review* (September) 82, 3: 875–85.

O'Connor, J. (1973) *The Fiscal Crisis of the State*, St Martin's Press, New York.

Offe, Claus (1972) 'Advanced capitalism and the welfare state', *Politics and Society* (Summer) 1, 4: 479–88.

Offe, Claus (1984) *Contradictions of the Welfare State* (ed. John Keane), Hutchinson, London.

Offe, Claus and Ronge, Volker (1975) 'Theses on the Theory of the State', *New German Critique* 6: 139–47.

Office of Population Censuses and Surveys (1981) *Census 1981: definitions*, HMSO, London.

Ollsson, Sven E. (1990) *Social policy and welfare state in Sweden*, Arkiv Förlag, Lund.

180   References

O'Toole, Richard and O'Toole, Anita Werner (1981) 'Negotiating inter-organizational order', *Sociological Quarterly* (Winter) 22, 1: 29–41.

Oxley, Michael (1989) 'Housing policy: comparing international comparisons', *Housing Studies* (April) 4, 2: 128–32.

Pahl, R.E. (1969) *Whose City?*, Penguin, London.

Pahl, R.E. (1977) 'Managers, technical experts and the state: forms of mediation and dominance in urban and regional development' in Harloe (1977, 49–60).

Parkinson, Michael (1989) 'The Thatcher government's urban policy, 1979–89', *Town Planning Review* (October) 60, 4: 421–40.

Pinker, Robert (1971) *Social Theory and Social Policy*, Heinemann, London.

Pinker, Robert, (1986) 'Social welfare in Japan and Britain', in Else Øyen (ed.) *Comparing Welfare States and their Futures*, Gower, Aldershot.

Piven, Francis Fox and Cloward, Richard A. (1972) *Regulating the Poor: the functions of public welfare*, Tavistock, London.

Popenoe, David (1977) *The suburban environment: Sweden and the United States*, University of Chicago Press, Chicago, IL.

Poulantzas, Nicol (1973) *Political Power and Social Classes*, New Left Books, London.

Rådberg, Johan (1988) *Doctrine and density in Swedish urban development 1875–1975* (in Swedish), Swedish Building Research Council Report (R11: 1988), Stockholm.

Rein, Martin, Esping-Andersen, Gøsta, and Rainwater, Lee (eds) (1987) *Stagnation and Renewal in Social Policy: the rise and fall of policy regimes*, M.E. Sharpe, New York.

Rex, John (1973) *Race, colonialism and the city*, Routledge, London.

Rex, John and Moore, Robert (1967) *Race, Community and Conflict*, Oxford University Press, Oxford.

Rokkan, Stein (1970) *Citizens, Elections, Parties: approaches to the comparative study of development*, Universitetsförlag, Oslo.

Rose, Richard (1986a) 'Common goals but different roles: the state's contribution to the welfare mix', pp. 13–39 in Rose and Shiratori (1986).

Rose, Richard (1986b) 'The dynamics of the welfare mix in Britain', pp. 80–106 in Rose and Shiratori (1986).

Rose, Richard and Shiratori, Rie (eds) (1986) *The Welfare State East and West*, Oxford University Press, Oxford.

Rostow, W.W. (1960) *The Stages of Economic Growth: a non-communist manifesto*, Harvard University Press, Harvard (Mass.).

Rothstein, Bo (1990) 'Marxism, institutional analysis, and working-class power: the Swedish case', *Politics and Society* (September) 18, 3: 317–45.

Ruesche, Georg and Kirchheimer, Otto (1939) *Punishment and Social Structure*, Russell & Russell, New York.

Ruonavaara, Hannu (1987) 'The Kemeny approach and the case of Finland', *Scandinavian Housing and Planning Research* (August) 4, 3: 163–7.

Rydin, Y. and Myerson, G. (1989) 'Explaining and interpreting ideological effects: a rhetorical approach to green belts', *Society and Space* 7: 463–79.

Safa, Helen Icken and Levitas, Gloria (eds) (1975) *Social Problems in Corporate America*, Harper and Row, New York.

Salamini, Leonardo (1981) *The Sociology of Political Praxis: an introduction to Gramsci's theory*, Routledge, London.

Printed in the United Kingdom by Lightning Source

Sarre, P. (1986) 'Choice and constraint in ethnic minority housing: a structurationist view', *Housing Studies* (April) 1, 2: 71–86.

Sassoon, Anne Showstack (1980) *Gramsci's Politics*, Croom Helm, London.

Sassoon, Anne Showstack (ed.) (1982) *Approaches to Gramsci*, Writers and Readers Publishing Cooperative Society, London.

Saunders, Peter (1978) 'Domestic property and social class', *International Journal of Urban and Regional Research* (June) 2, 2: 233–51.

Saunders, Peter (1979) *Urban Politics: a sociological interpretation*, Hutchinson, London.

Saunders, Peter (1986) *Social Theory and the Urban Question*, Hutchinson, London.

Saunders, Peter (1990) *A Nation of Home Owners*, Unwin Hyman, London.

Saunders, Peter and Williams, Peter (1988) 'The constitution of the home: towards a research agenda', *Housing Studies* (April) 3, 2: 81–93.

Scanzoni, John (1979) 'The centrality of negotiation in the study of social organization', *Contemporary Sociology* 8: 528–30.

Schmidt, Stephan (1989a) 'Convergence Theory, Labour Movements, and Corporatism: the case of housing', *Scandinavian Housing and Planning Research* (May) 6, 2: 83–101.

Schmidt, Stephan (1989b) 'Review of Ball *et al.* (1988)' *Scandinavian Housing and Planning Research* (February) 6, 1: 60–2.

Simon, Roger (1982) *Gramsci's Political Thought: an introduction*, Lawrence & Wishart, London.

Skocpol, Theda (1979) *States and Social Revolutions: a comparative analysis of France, Russia and China*, Cambridge University Press, Cambridge.

Smith, Michael Peter (1980) *The city and social theory*, Basil Blackwell, Oxford.

Somerville, Peter (1988) 'Home sweet home: a critical comment on Saunders and Williams', *Housing Studies* (April) 4, 2: 113–19.

SOU (1933) *Investigation concerning the need to augment housing statistics* (in Swedish), 1933: 14, Stockholm.

SOU (1945) *Final Report of the Social Housing Investigation: Part I, general guidelines for future housing policy: proposals for loan and grant systems* (in Swedish), 1945: 63, Stockholm.

SOU (1952) *Homehelp: Collective Residence Committee Report I* (in Swedish), 1952: 38, Stockholm.

SOU (1954) *Collective Housing: Collective Residence Committee Report II* (in Swedish), 1954: 3, Stockholm.

SOU (1955a) *Laundering: Collective Residence Committee Report III* (in Swedish), 1955: 8, Stockholm.

SOU (1955b) *Public Meeting Facilities: Collective Residence Committee Report IV* (in Swedish), 1955: 28, Stockholm.

SOU (1956) *The Home and Societal Planning: Final Report of the Collective Residence Committee* (in Swedish), 1956: 32, Stockholm.

SOU (1974) *A solidaristic housing policy* (in Swedish), 1974: 17, Stockholm.

SOU (1990) *Commission of Investigation into Cities* (in Swedish), 1990: 20, Stockholm.

Spector, Malcolm and Kitsuse, John I. (1973) 'Social Problems: a reformulation', *Social Problems* (Fall) 21, 2: 145–59.

Spector, Malcolm and Kitsuse, John I. (1977) *Constructing Social Problems*, Cummings, Menlo Park, CA.

Stephens, J. (1979) *The Transition from Capitalism to Socialism*, Macmillan, London.

Stockholm (1945) *The Future Stockholm: guidelines for Stockholm's General Plan*, Stockholm Town Planning Office (in Swedish).

Strauss, Anselm L. (1978) *Negotiations: varieties, contexts, processes and social order*, Jossey-Bass, San Francisco, CA.

Strauss, Anselm L. (1982) 'Interorganizational negotiation', *Urban Life* (October) II (3), 350–67.

Strauss, A.L., Schatzman, L., Ehrlich, D., Bucher, R. and Sabshin, M. (1963) 'The hospital and its negotiated order', in Eliot Friedson (ed.) *The Hospital in Modern Society*, Free Press, New York 147–69.

Strauss, A.L., Schatzman, L., Ehrlich, D., Bucher, R. and Sabshin, M. (1964) *Psychiatric Ideologies and Institutions*, Free Press, New York.

Szalai, Alexander (1972) *The use of time: daily activities of urban and suburban populations in twelve countries*, Mouton, The Hague.

Szelenyi, Ivan (1989) 'Housing policy in the emergent socialist mixed economy of Eastern Europe', *Housing Studies* (July) 4, 3: 167–76.

Szelenyi, Ivan and Manchin, Robert (1987) 'Social policy under state socialism: market redistribution and social inequalities in East European socialist societies', pp. 102–39 in Rein *et al.* (1987).

Taylor-Gooby, Peter and Dale, Jennifer (1981) *Social Theory and Social Welfare*, Edward Arnold, London.

Thrift, Nigel (1983) 'On the determination of social action in space and time', *Society and Space* 1: 23–57.

Tilton, Tim (1988) 'The role of ideology in social democratic policy' pp. 369–88 in K. Misgeld, K. Molin, and K. Åmark (eds) *Social democracy's society: the SAP and Sweden over 100 years* (in Swedish), Tiden, Stockholm.

Todd, Jean and Griffiths, David (1986) *Changing the definition of a household*, Office of Population Censuses and Surveys, Social Service Division, HMSO, London.

Torgersen, Ulf (1987) 'Housing: the wobbly pillar under the welfare state', in Bengt Turner, Jim Kemeny, and Lennart J. Lundqvist (eds) *Between State and Market: housing in the postindustrial era*, Almqvist & Wiksell, Stockholm.

Turner, Bryan S. (1987) *Medical Power and Social Knowledge*, Sage, London.

Turner, Bryan S. (1990) 'The interdisciplinary curriculum: from social medicine to postmodernism', *Sociology of Health and Illness* (March) 12, 1: 1–23.

Urry, John (1981) *The Anatomy of Capitalist Societies: the economy, civil society and the state*, Macmillan, London.

Useem, Bert and Zald, Mayer N. (1982) 'From Pressure Group to Social Movement: organisational dilemmas of the effort to promote nuclear power', *Social Problems* (December) 30, 2: 144–56.

Vidich, Arthur J. and Bensman, Joseph (1958) *Small Town in Mass Society: class, power and religion in a rural community*, Princeton University Press, Princeton, NJ.

Wallerstein, Immanuel (1984) 'The development of the concept of development' pp. 102–16 in Randall Collins (ed.) *Sociological Theory 1984*, Jossey-Bass, San Francisco, CA.

Wildavsky, Aaron (1987) 'Choosing preferences by constructing institutions: a cultural theory of preference formation', *American Political Science Review* (March) 81, 1: 3–21.

Wilensky, Harold L. (1975) *The Welfare State and Equality: structural and ideological roots of public expenditure*, University of California Press, Berkeley, CA.

Wilensky, Harold L. (1981) 'Leftism, catholicism, and democratic corporatism: the role of political parties in recent welfare state development', in P. Flora and A.J. Heidenheimer (eds) *The Development of Welfare States in Europe and America*, Transaction Books, New Brunswick, NJ.

Wilensky, H.L., Luebbert, G.M., Hahn, S.R. and Jamieson, A.M. (1985) *Comparative Social Policy: theories, methods, findings*, Institute of International Studies, Berkeley, CA.

Williams, Gwyn (1960) 'Gramsci's concept of 'egemonia'' *Journal of the History of Ideas* No. 4.

Wilson, Dorothy (1979) *The welfare state in Sweden: a study in comparative social administration*, Heinemann, London.

Wrong, Dennis (1961) 'The oversocialized conception of man in modern sociology', *American Sociological Review* (April) 26, 2: 183–93.

Index